CONTENTS

WALKDEN YARD

WALKDEN YARD

YARD

THE LANCASHIRE CENTRAL COALFIELD WORKSHOPS

ALAN DAVIES

AMBERLEY

First edition 2013

Amberley Publishing
The Hill, Stroud
Gloucestershire, GL5 4EP

www.amberleybooks.com

British Library Cataloguing in Publication Data.
A catalogue record for this book is available from the British Library.

ISBN 978-1-84868-925-1

Typeset in 10pt on 12pt Sabon.
Typesetting and Origination by Amberley Publishing.
Printed in the UK.

Acknowledgements

Thanks to the late Joe Cunliffe (died September 1993), former superintendent at Walkden Yard, for allowing me to copy in the 1990s many of the photographs and loco maintenance documents formerly held there. Joe was very enthusiastic about the idea of a publication; he also warned me that there were many enthusiasts out there who had amassed huge amounts of information about Walkden Yard, more than he himself could ever remember! Far more information surfaced than was expected so it was decided to follow this publication with one purely devoted to the locomotives and collieries served by Walkden Yard.

Philip Hindley, railway historian and photographer, has been extremely generous and helpful in ensuring that facts, figures and technical information were accurate. His meticulous attention to detail has inspired me to try that bit harder with the accuracy of my research. Thanks to Glen Atkinson, long standing historian of the Duke of Bridgewater's underground and surface canals along with the Walkden area, for allowing free access to his research and local history brain cells. Many thanks to Bessie Hilton who in her ninety-first year allowed me to interview her and copy photographs and documents relating to her father Fred Hilton, manager at Walkden Yard from 1924 until 1950.

Thanks over the years to local historians Carol Woodward of Boothstown and the late and sadly-missed bundle of energy Walkden librarian Anne Monaghan who left us far too young. Thanks also to the late Frank Mullineux and Elsie Mullineux for their work over the years and enthusiasm for all things Walkden and Worsley. Thanks as ever for all things railway related to Dennis Sweeney of Leigh who freely gives of his extensive knowledge.

Thanks to Roger Fielding for the use of his photographs. As hard as I tried I could not track down George Booth or his family but relied on those who worked on his publications back in the 1980s, who said that he would have been very excited at the prospect of a book about Walkden Yard including extracts from his reminiscences.

Norman Glyn of Tyldesley, former employee at Walkden Yard, for his reminiscences. Mick Hughes of Atherton for his reminiscences of footplate work in the 1960s. Charles Birdsall for reproducing his rare Walkden Yard pay tally. Tim Boddington for his background information on the Burrows family.

The research carried out since at least the 1930s by (later to be) members of the Industrial Locomotive Society goes without saying as regards its historical importance today and in the future. The work of the society's enthusiastic members (some now sadly deceased) – including C. Harry Townley, C. Alex Appleton, Frank D. Smith, Jim Peden, Cyril Golding, Cliff Shepherd, Steven Oakden and Philip G. Hindley – has meant a detailed record will survive for future generations of an era never to be seen again. I am very grateful that the Society allowed me to quote from their articles and to use the superb plans of the rail network by Ian Lloyd, which greatly enhance this work.

Thanks especially to Steve Leyland of Bolton for access to and the use of his recollections and images from his historically-important photographic archive. Thanks also to graphic designer par excellence Neil Duerden (www. neilduerden.co.uk) for his image-manipulation mastery. Thanks also to Anthony Coulls, Senior Curator at the National Railway Museum, for information relating to *Princess*, NSR/LMS No. 2. Many thanks to Jack de Guillaume and daughter Wendy Williams for his Walkden Yard motor body repair shop photo. Also coming in at the last minute was ex-Walkden man Brian Wharmby with his reminiscences and fascinating photographs – again many thanks. Thanks as ever to Bruce Jackson, County Archivist at the Lancashire Archives in Preston for allowing reproduction of the Bridgewater Collieries' records.

Preface

For those who lived during the generations when Walkden Yard workshops were active – for the men who worked there, their families, those regularly passing by, or those hearing about it in local conversation – the site had an especially familiar ring to it for miles around. It was as though Walkden, usually termed 'Wogdin' Yard locally, was a community in itself rather than an engineering workshop similar to any other. I remember seeing a plan, possibly mid-eighteenth-century pre-Bridgewater Canal, which spelt Walkden as Waugden, hence the long-standing corruption of what is probably an ancient place name. Walkden Yard lay to the south of High Street, adjacent on its east to Ellesmere Colliery. Despite the name Walkden Yard, a large portion of the yard actually lay in the township of Little Hulton.

In his reminiscences of the period 1932–1939 (published in 1988) former employee George Booth recalled,

> The name 'Walkden Yard' must conjure up in the minds of the older inhabitants of Walkden and its surrounding district a regular 7.00 am hooter; a hot flush of workmen ascending Tynesbank to catch trams home (or to the pub); hustling, snorting and steaming locomotives straining 'up the bank' to Ashton Field; and the faithful chiming of 13 by 'Lady Bourke' [the copy of the famous late-eighteenth-century Bridgewater Trustees clock at Worsley Yard, later the Manchester Collieries, then NCB headquarters clock at Walkden, which purposely struck thirteen at one o'clock].

The old Bridgewater Trustees mineral railways were to become the Central Railways of the huge Manchester Collieries concern formed in March 1929. The landscape with its changing, suddenly abrupt and often fierce gradients was to be a cruel one for these colliery locomotives which were almost constantly worked to their limits. From Worsley to Linnyshaw Colliery east of Walkden the average gradient had been 1 in 52 with the occasional 1 in 30 stretch! In later days a typical run might be from Astley Green Colliery, south-west of Walkden on the 98-foot contour to Mosley Common Colliery at 177 feet, Walkden Yard at 291 feet, Ashton's Field 360 feet, with the branch to Brackley Colliery ending up at 374 feet.

To summarise the system (courtesy of the Industrial Railway Society): from the west we have BR at Astley Moss Sidings heading north to Astley Green Colliery and loco shed (1¾ miles). The line headed east to Boothsbank Tip canal wharf (locally known as Boothstown Basin; 3 miles) where coal was tipped into barges. A northerly change of direction brought the line to the important Mosley Common Colliery (3½ miles), beyond which connection was made with BR west of Ellenbrook station. Continuing north-east to Walkden led to connections with two BR lines: Ellesmere Sidings, ½ mile west of Walkden High Level station, and Walkden Sidings, ¼ m west of Walkden Low Level station. Next stop was Walkden Yard (4¾ miles) with loco shed. North-west of Walkden Yard was Ashton's Field Colliery (5¾m) with coal-blending plant and landsale depot (Buckley Lane depot).

From Ashton's Field a line ran east for ¾ mile to a junction. A branch turned north-east to a connection at Linnyshaw Moss (¼ mile) with the BR Kearsley branch. Beyond this point the line from Ashton's Field headed south-east and later south to Sandhole (Bridgewater) Colliery with its washery and loco shed (2 miles). South of Sandhole led to a BR connection at Sandersons Sidings ½ mile north-west of Worsley station. The line then continued to Worsley Tip canal wharf (3½ miles). North of Sandhole Colliery, heading north-east then east, was a line to the Moss Lane landsale depot, Pendlebury (2 miles).

From Ashton's Field Colliery a line ran north-west to a junction with a line north-east to Dixon Green (Farnworth) landsale depot (½ mile). The line carried on west to Brackley Colliery and loco shed (1¾ miles) and the ever-growing Cutacre waste tips (2 miles), which are being worked and landscaped as I write. A connection with the BR Little Hulton branch was finally made (2¼ miles).

As the older collieries gradually closed, the Moss Lane branch closed in 1956, Sandersons Sidings–Worsley Canal closed around 1961, the Sandhole Colliery–Sandersons Sidings line closed in 1966. The Dixon Green branch line closed in 1966, the Ashton's Field Colliery–Brackley Colliery closed in February 1968. The Linnyshaw Moss to Sandhole Colliery line closed in September 1968. The last lines from Ashton's Field Colliery to Linnyshaw Moss and the Astley Moss sidings closed in October 1970.

The staff at Walkden Yard worked wonders, constantly overhauling these hard-working locomotives. They had the skills to tackle virtually any job required with the benefit of generations of experience handed down, many times from father to son.

Not all the locos which passed through Walkden Yard feature in this work, which is mainly about the Yard itself. They will feature in a second study to follow this one.

In no way is this a definitive history of Walkden Yard and the surrounding rail system; the types of archive to produce such a work have not survived. It relies heavily on the Townley, Smith, Peden, Appleton 1995 study and the meticulous research of the Industrial Railway Society along with many knowledgeable individuals, but at least it gives a broader picture including aspects of social history for the general reader.

Being a coal-mining historian I realise I have stepped into very dangerous territory: that of the hallowed ground of the railway historian, that most famous of breeds that Britain is rightly famous for! As such I have not attempted detailed wider interpretation of images and maintenance information, but have merely supplied them. It has also been made clear to me that there are dangers in trying to document the movements of the various locomotives over their lives and that there are occasional conflicts among rail historians in this respect. To try and resolve these would be impossible due to the lack of an archive record documenting all the movements of the locomotives.

Alan Davies
Tyldesley
2013

The Worsley and Walkden Workshops and Rail Network: Historical Background

Worsley Yard, Precursor to Walkden Yard

A visitor today to the idyllic Worsley Green, south of the East Lancashire Road and hemmed in amongst a motorway hell, with its exclusively-priced housing, could never imagine how purely industrial an area it had once been.

Coal mining in the Worsley and Walkden area is truly ancient – documented in archives from 1376, undocumented mining activity no doubt existed far earlier. Leases for organised and regular coal extraction begin to appear from the early years of the sixteenth century. The seventeenth century brought few major changes to the techniques of mining at shallow depths, as the burden of water and gas removal was a major restriction on expansion.

As nearby Manchester rapidly expanded with an increased demand for coal, yet was hampered by poor transport links, Francis Egerton 3rd Duke of Bridgewater, recognised his opportunity and proposed the construction of a surface canal from Booth's Bank to Worsley. An Act came into force in 1759, with an extension reaching Manchester by 1765.

At Worsley Delph (an old sandstone quarry) an underground canal or navigable level was driven north-westward towards Walkden and Farnworth. Two great geniuses/engineers of their time, John Gilbert and James Brindley, both of whom had developed their skills in mining districts in Derbyshire and Staffordshire, masterminded the project.

The canal was passing below Walkden by 1770, and below Buckley Lane (near Ashton's Field, Little Hulton) by 1801. For many years, over 100,000 tons of coal a year were produced and transported to Worsley Delph using the main ground-level canal, this system being in operation until 1887.

The Duke of Bridgewater had gradually established a workshop at Worsley Green from around 1759, servicing his surface-canal construction and also the continued drivage of the mine drainage tunnel, which was later to become part of the vast underground canal transport system for the network of

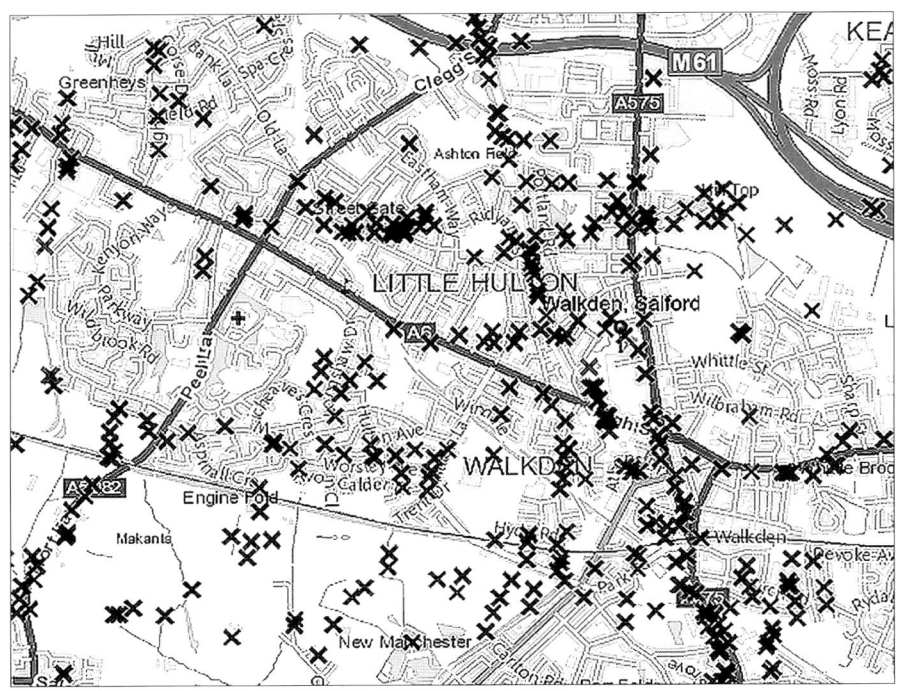

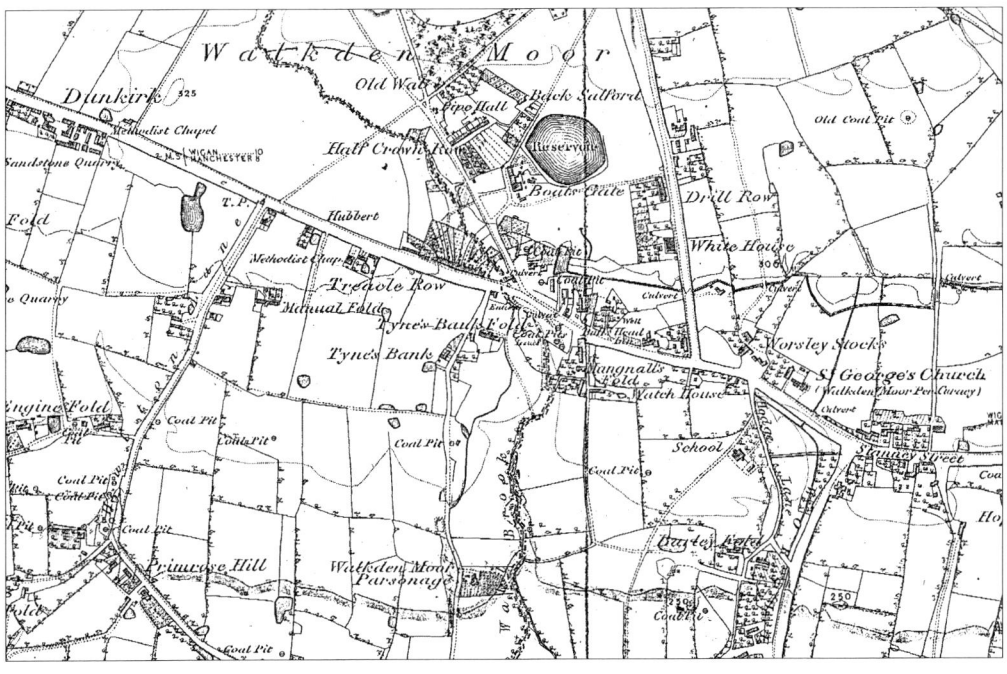

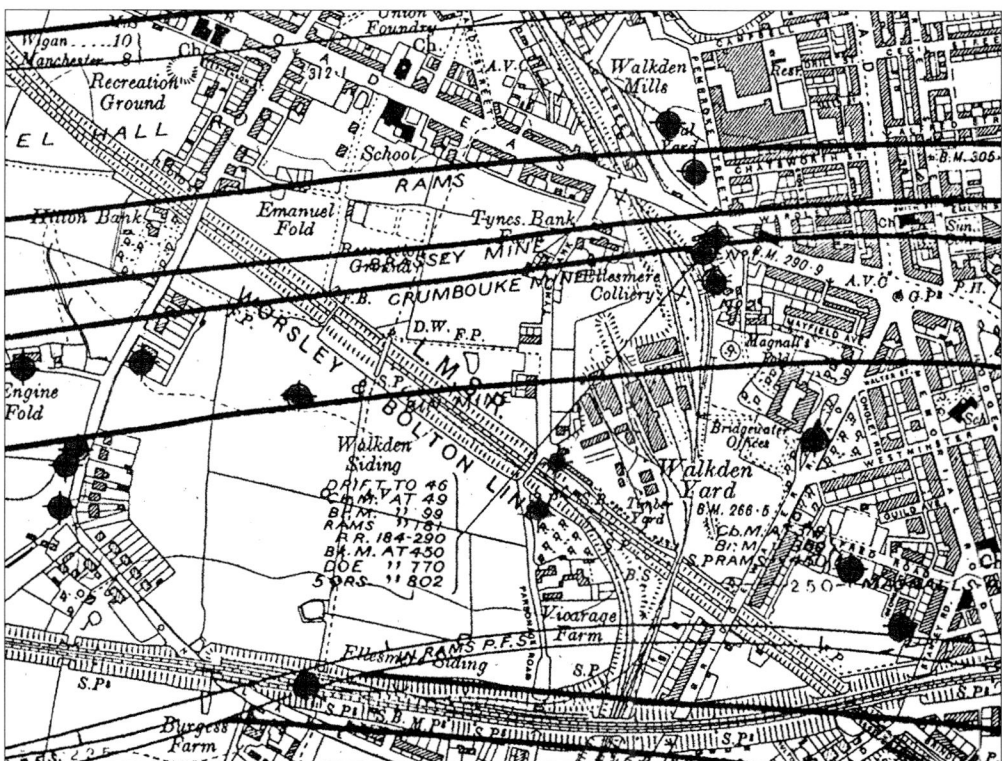

The 1929 edition of the 6-inch Geological Survey sheet 95SW shows coal seams outcropping across central Walkden, notably the Rams, Brassey and Crombouke, the seams dipping from the surface to the south-east at about 1 in 5. Walkden Yard is seen middle centre, with the adjacent Ellesmere Colliery accessing the Five Quarters seam at 802 feet.

Opposite above: The Coal Authority shaft register gives an idea of the extent of coal mining activity in the Little Hulton to Walkden area over the centuries. Many more unrecorded shafts dating from the late sixteenth to early nineteenth centuries will be present, revealing themselves unexpectedly, no doubt, to future generations.

Opposite below: Sheet 95 of the First Edition 1850 OS series shows the large reservoir at Boats Gate, north of Walkden town centre. The keener-eyed will notice many shafts dotted around nearby, mostly accessing the navigable level and its branches. Immediately south of the reservoir would be the site of the first Walkden Workshops. South of the future A6, to the right of Tyne's Bank, would be the site of Walkden Yard sixty years later.

A fascinating early view of Worsley Delph in use as a sandstone quarry. The single drainage level with doors to the right was to become the eastern arm of two navigable entrances to the vast underground canal system.

collieries. Colliery mechanical engineers, miners (male and female until 1842), labourers, carpenters and boat builders along with lime burners and mortar makers populated Worsley Green. Worsley Brook supplied a mill dam behind Mill Brow with water to drive a 24-foot-diameter waterwheel. This ground corn for the duke's estate and workforce as well as serving as a limestone mortar mill, providing mortar for the huge number of bricks required to line the main underground canal tunnels.

Below Walkden and Farnworth lay the upper navigable level, this canal being 30 yards above the main tunnel. Being so close to the surface the obvious thing to do was to drive an inclined drift or tunnel from the canal below ground to the surface, enabling new barges to be sent into the system or others removed for maintenance. The upper navigable level was driven around 1773, created by widening the original Massey drainage sough with a finished length of 1.75 miles (2.8 km) and draining into the main navigable level. The inclined drift from the upper canal level to the surface, at a gradient of 1 in 4, allowed boats to be drawn up and lowered down. This incline had been continued downwards to the main level. Further navigable levels were dug below the main level to serve deeper seams; these were 57 yards (52 m) and 83 yards (76 m) deeper. The upper canal had gone into disuse as regards coal haulage by 1820.

Boats Gate Yard Workshops, Walkden

From 1773 to 1780 the driving of the upper canal level and inclined tunnel to the surface at Walkden was completed. Gradually boat workshops and a reservoir known as Boats Gate (1787) were constructed close by, just to the north of Campbell Way near the present-day Tesco petrol station (2013). The reservoir at Walkden had a triple purpose: to 'top up' the upper canal level and subsequently those below as water naturally flowed out, to store barges being built or repaired and also to supply water to colliery water-bucket gravity winding engines nearby.

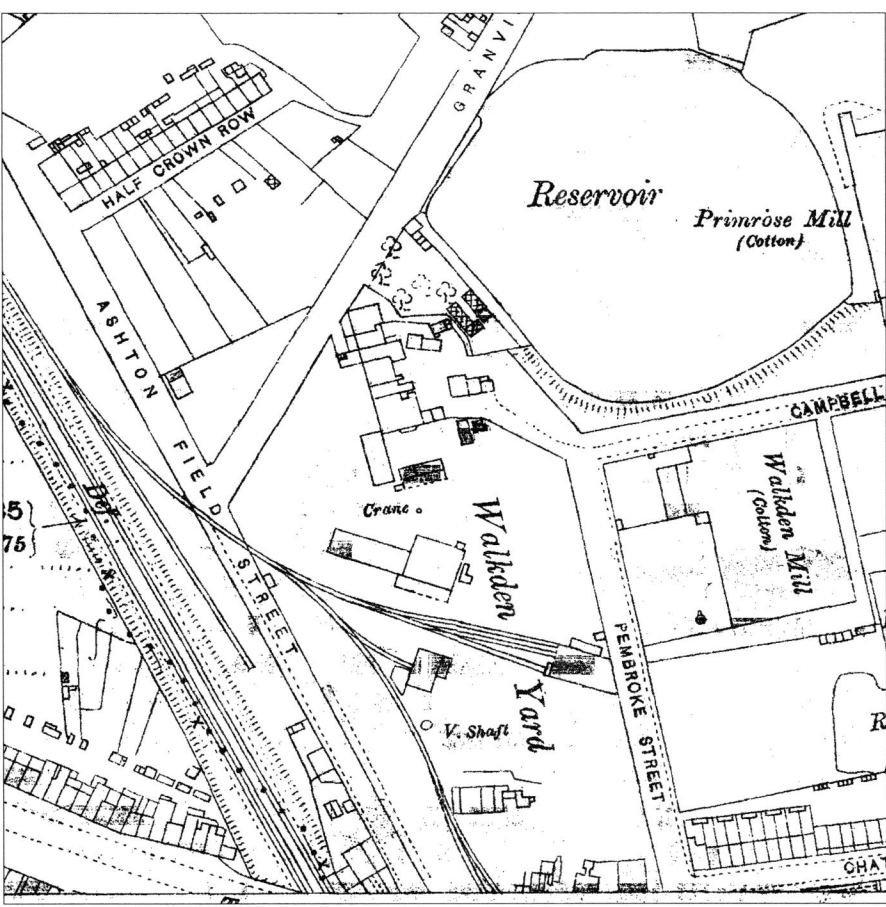

The first Walkden Yard workshops, completed by approximately 1778, can be seen north of the A6 (where today's Tesco car park is situated) on the 1893 Sheet 95.10 25-inch OS plan. Boatshed reservoir is top right, topping up the upper and main navigable levels. An inclined drift tunnel was situated close to the crane, allowing barges to be brought to surface for repair. Mineworkers housing is seen top left at 'Half Crown Row'.

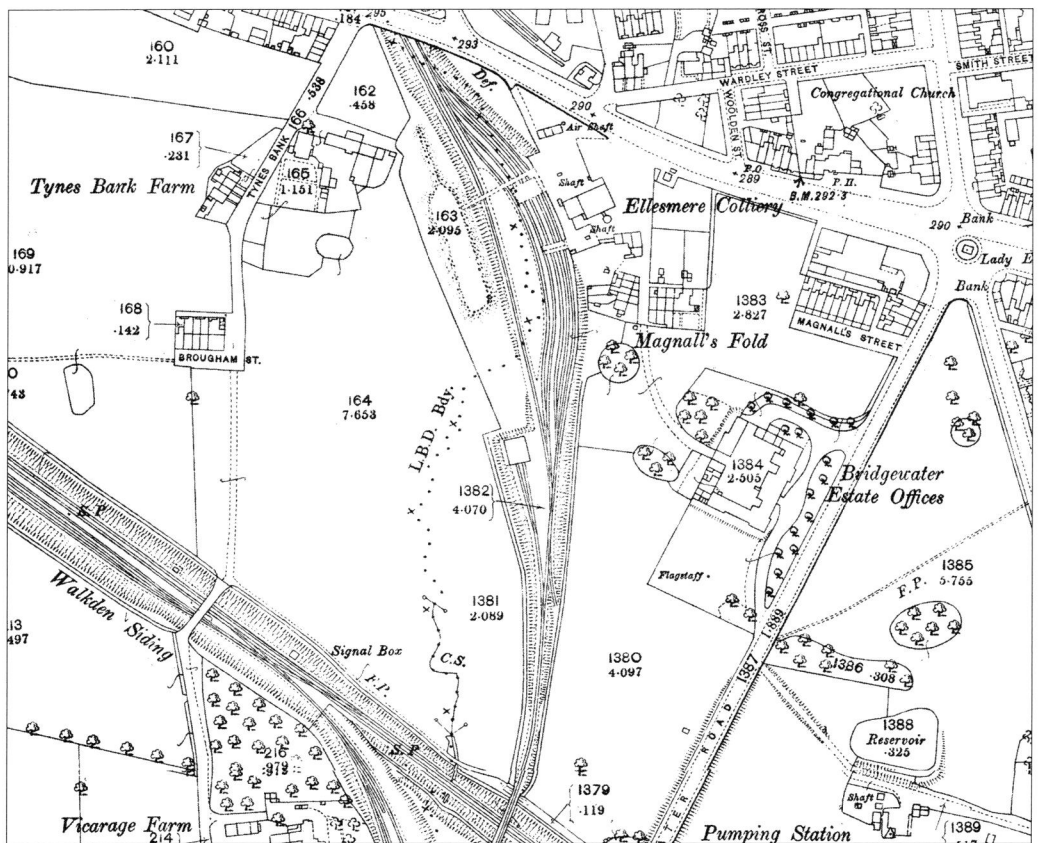

A closer look at Sheet 95.14, the 25-inch OS plan of 1893. The site of the second Walkden Yard is still farmland west of Ellesmere Colliery. The substantial Bridgewater Estate offices built in 1868 stand to the right.

Lime kilns, a coal landsale area and coke ovens were built here between the reservoir and High Street. A colliery basket and chain workshop also stood nearby, the maintenance complex being mostly completed by 1778. This workshop was to become the first 'Walkden Yard'.

After the Third Duke, 1803

The 3rd Duke of Bridgewater died on 8 March 1803. His industrial properties, along with his estates in Lancashire and Brackley in Northamptonshire, were separated from his other estates in Shropshire and Hertfordshire and placed in trust for the benefit of the duke's nephew, Earl George Gower, later the Duke of Sutherland. By the terms of the will, control was vested in a body of three –

the Trustees of the Duke of Bridgewater. This arrangement lasted for a century (the trustees sold the Bridgewater Canal to the Bridgewater Navigation Co. Ltd in 1872).

An Early Walkden Foreman

In March 1843 Daniel Timmins was the foreman engineer at this first Walkden Yard workshop/stores. His draft revised contract/job description for March 1843 has survived in the records held at the Lancashire Record Office:

> March 25th 1843
> Directions to be observed by the foreman (Daniel Timmins) foreman of the engines and engineers.
> 1. To have charge of all cast iron for the Engines, Pits & Machinery except castings for wagons & wagons roads.
> 2. To keep charge of all Blocks & block ropes, capstan ropes and turn ropes for the Engine Pits (the different inspectors to have charge of coal ropes, gin ropes & water ropes).
> 3. To make orders to the Colliery Agent for all Engine Stores, and have charge of the same, also to see that the same are properly received as per the demands & invoices.
> 4. To report to Mr Fereday Smith once a fortnight the state of repairs on hand also what other repairs are most wanted.
> 5. To make demands for all work, men and materials wanted from the Colliery Yard with proper description of the said work and if wanted the said demands to be attested by the colliery agent except for repairs where the articles are to be used again at the Engine Pit or colliery from which they came from.
> 6. In cases of emergency an order may be given direct to Will<u>m</u> Roscow or in his absence to his deputies.
> 7. Time made by the Carpenters, Blacksmiths or Boiler men at the Collieries, Engines or Pits away from the shafts to be kept by the Inspectors of the said pits, the work to be under the direction of the Engineer or the Foreman of the department for which they are working.

As the vast Bridgewater mining system expanded from the late eighteenth century onwards the collieries closer to Worsley had utilised horse tram roads. An underground barge inclined plane, horse and, later, stationary-engine tram roads linked a number of the collieries west and east of Worsley. To the west the 1¼-mile-long Ellenbrook tramway served Madams Wood Colliery

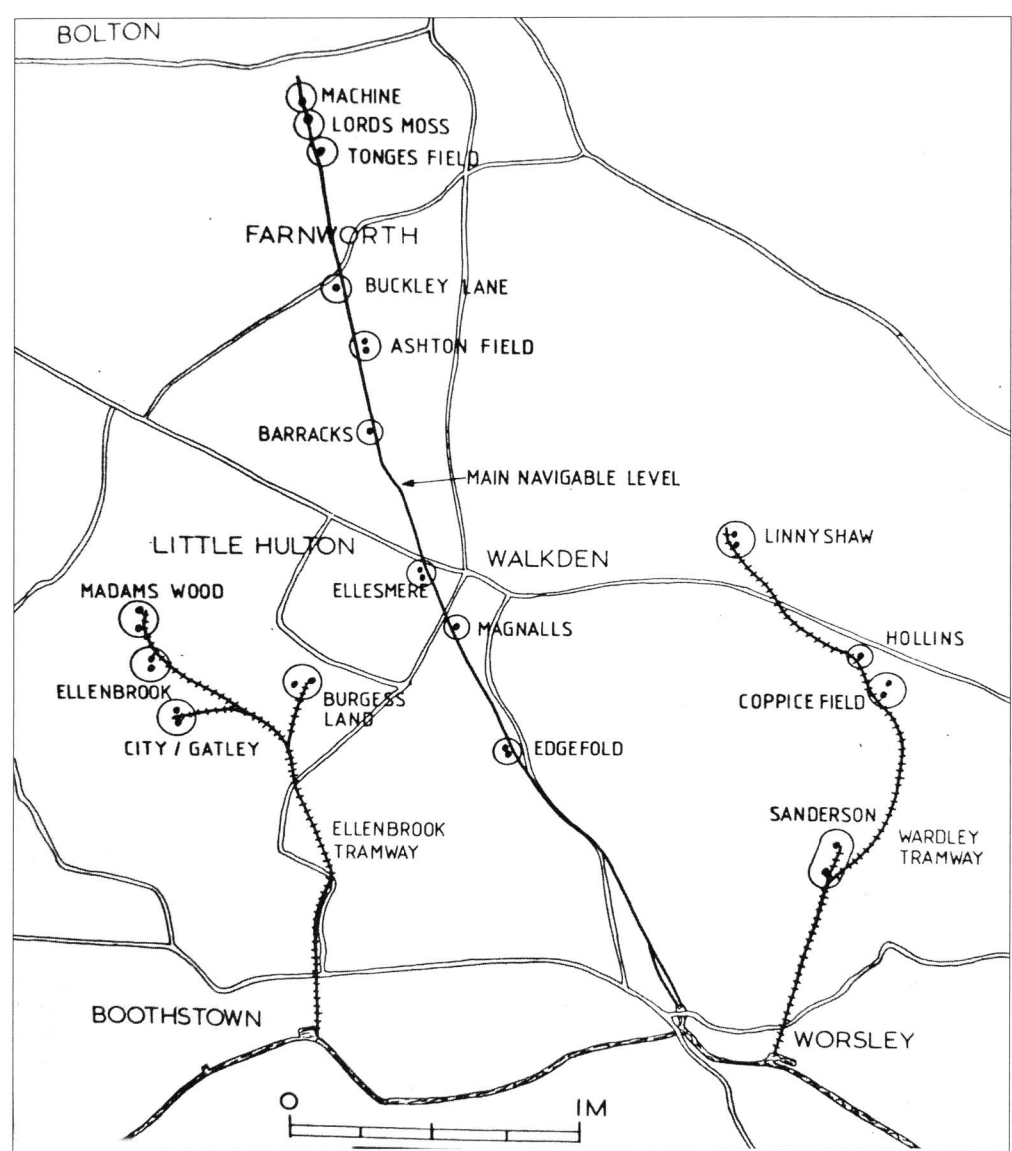

The old Bridgewater Collieries network of horse drawn tramways feeding the Bridgewater Canal coal drops at Boothstown and Worsley to the south. This plan covers the period 1860 to 1870, the main navigable level by then over three miles in length heading towards Bolton. (Plan courtesy Glen Atkinson)

Sketch shewing the method adopted by the Bridgewater Trustees of conveying coal from their various pits to their Canal, for shipment to Manchester & district.
Mar: 1863.

Two to three tons of coal heading to the canal by gravity on the Bridgewater 4ft gauge tramway, the wagon man applying the brake to control speed. The horse is having a temporary breather before what must have been very hard work returning the empty wagon up to two miles to the north.

and others, over 1 km to the north to the Bridgewater Canal. To the east the eventual 2¼-mile Wardley tramway linked Linnyshaw Colliery and other pits to Worsley, Linnyshaw being approximately 2 km to the north. These early tramways were still in use between the 1860s and 70s with 3-ton wagons individually braked.

The tramway system had been adequate up to a point, but would hamper long-term expansion of output. The arrival and widespread use of railways in many Lancashire colliery districts eventually led to the decision to link the Bridgewater system with the London & North Western Railway. A line was built through Eccles, Worsley and Wigan, opening in 1864.

The Bridgewater Trust then began to link their larger long-life collieries, Mosley Common and Sandhole, to the LNWR via two standard gauge lines meeting to the north and at the canal in the south. In 1865 John Claridge of Swinton took on the contract for the Buckley Lane, Ashton's Field Colliery and Linnyshaw Colliery line followed by a link to Sandhole Colliery. The first locomotives eventually arrived at Sandhole Colliery in July and September 1870 being hauled at the end of their journey by teams of horses due to the rail link not having been completed.

New Headquarters and Railway Expansion

A prestigious, new office building for the Trustees of the Duke of Bridgewater (locally named Bridgewater Offices) had been completed at Walkden in 1868. Here the Bridgewater Trust superintendent, general manager and colliery agent were based, along with surveyors and sales staff. Walkden took on a new level of importance as a central base for the vast mining network.

The substantial Bridgewater Trustees Estates offices at Walkden, built in 1868, are seen here around 1910, the Lady Bourke Clock having been added in December 1900. Mrs Bourke, wife of Walter Langley Bourke, Superintendent of the Bridgewater Trust, was not a 'lady' by title, although Bourke succeeded later to the title of 4th Earl Mayo.

The original famous thirteen-chiming clock was first placed in the tower at Worsley Yard in around 1780 to alert workmen to the beginning or end of their shifts. They claimed they either had or were in danger of missing it striking once at 1 p.m., so the clock was adapted to strike thirteen times at 1 p.m. When Worsley Yard finally closed in 1901, the clock was first removed to the gatehouse of Worsley New Hall in 1900–1, then moved to Bridgewater House, London, in around 1924, then given to St Mark's church, Worsley, on its centenary in 1944, where it continues in operation to this day.

The new Lady Bourke was a completely new clock, added to the Bridgewater Trust offices to celebrate the start of the twentieth century on 1 January 1901. It was electrically regulated and wound and designed to strike thirteen at 1 p.m. to continue the tradition. On New Year's Eve in 1954, its 400 lb drive weight plummeted 60 feet to the ground, the clock needing no repairs.

The building was the headquarters of Manchester Collieries from 1929 until the nationalisation of the industry in 1947, and was then the offices of the National Coal Board. It was demolished in 1976 for housing. The new Lady Bourke clock then apparently went into storage in one of the Salford Museums, and its present whereabouts are unknown.

From 1870 to 1871 the line from Boothsbank on the Bridgewater Canal to Mosley Common Colliery was begun. This line was linked to the LNWR at Ellenbrook near Boothstown with an extension to Ellesmere Colliery at Walkden opened in October 1871 (Ellesmere Colliery was close to and south of the A6 opposite the Walkden Tesco car park). The final section, from Ellesmere Colliery to Ashton's Field Colliery to the west at Little Hulton/Farnworth, was tendered in July 1872. With the opening of the LNWR branch from Roe Green to Little Hulton a link was made with Ellesmere Colliery. Finally a connection was made from the Linnyshaw Colliery line to the Lancashire & Yorkshire Railway's new Kearsley mineral railway in the late 1870s.

With the completion of the trustees' own railway network in the 1870s it was announced in the *Eccles Advertiser* on 14 May 1870 that the first locomotive had arrived, although not mentioning its identity.

The sale of the Bridgewater Canal in 1872 and the decreasing use of the underground canal system for coal transport now meant that the company's maintenance and workshop infrastructure needed to be reassessed and upgraded. The line to Mosley Common Colliery was completed by 1873 along with further additions to the expanding fleet of wagons. An early locomotive arrival that we do have a record of was the Manning Wardle 'Grand Duchess' purchased in 1874.

Magnalls Colliery, also known as Rough Field Pit, is seen here around 1910. It was situated close to Ellesmere Colliery and to the north of the 1868 Bridgewater Offices. An example of an early beam pumping engine station of around 1840, its wooden headgear was used both for shaft access and pump spear rod maintenance. It closed as a working colliery by 1885, and its shaft was filled in 1912.

Further colliery sinkings, along with closures of the older pits, continued from the 1860s onwards. In 1878 sinking of Brackley Colliery in Little Hulton began and was to be a feature of the industrial landscape until May 1964. Sinking of additional shafts at Mosley Common Colliery took place in 1881 requiring substantial capital outlay. This colliery was to expand enormously in NCB days becoming one of the largest collieries in the British coalfields, surviving until February 1968. One victim of exhaustion of reserves in the 1880s was the old Magnalls Pit, which closed by 1885, situated very close to the future Walkden Yard and Bridgewater Offices (see photo).

The combination of major new colliery sinkings and the gradual creation of a mineral rail network allowed for a massive expansion in coal output, quadrupling from 1860 to 1886 (see table).

Year	Tons	Year	Tons
1803	134,035	1845	279,212
1829	153,630	1846	307,649
1830	165,634	1847	270,084
1831	178,161	1848	254,030
1832	174,260	1849	281,822
1833	220,369	1850	278,713
1834	229,421	1851	n/a
1835	271,840	1852	274,380
1836	297,276	1853	n/a
1837	273,225	1854	n/a
1838	n/a	1855	279,140
1839	308,660	1860	371,408
1840	283,709	1871	*c.* 550,000
1841	254,117	1886	1,303,625
1842	n/a		
1843	n/a		
1844	282,678		

By 1886 the Bridgewater collieries were regularly raising around 1,300,000 tons of coal per annum. In 1887 coal haulage using the underground canal system ended, with the links to the LNWR and L&YR and the Worsley canal loading point handling outgoing coal traffic.

Joseph Hyde's recollections of the old Worsley Yard and village in 1887 give us an idea of the size and range of the workforce ten years before work began on Walkden Yard. The excerpt was published in Manchester Collieries'

A poor quality but early view of Worsley Yard from the south canal side of the site in about 1870. On the clearance of Worsley Yard in around 1905, the chimney base was retained as part of a memorial to the Canal Duke and survives today on Worsley Green, now ironically a select residential area.

Worsley Yard in about 1880. The four-plank wagon has the latest spring buffers, painted with details of a string of sidings empties should be returned to, those legible being Sandersons Sidings and Ellesmere Sidings – the other is possibly Lynnyshaw Sidings. The stacked timbers are 'half bars', used in roof support below ground.

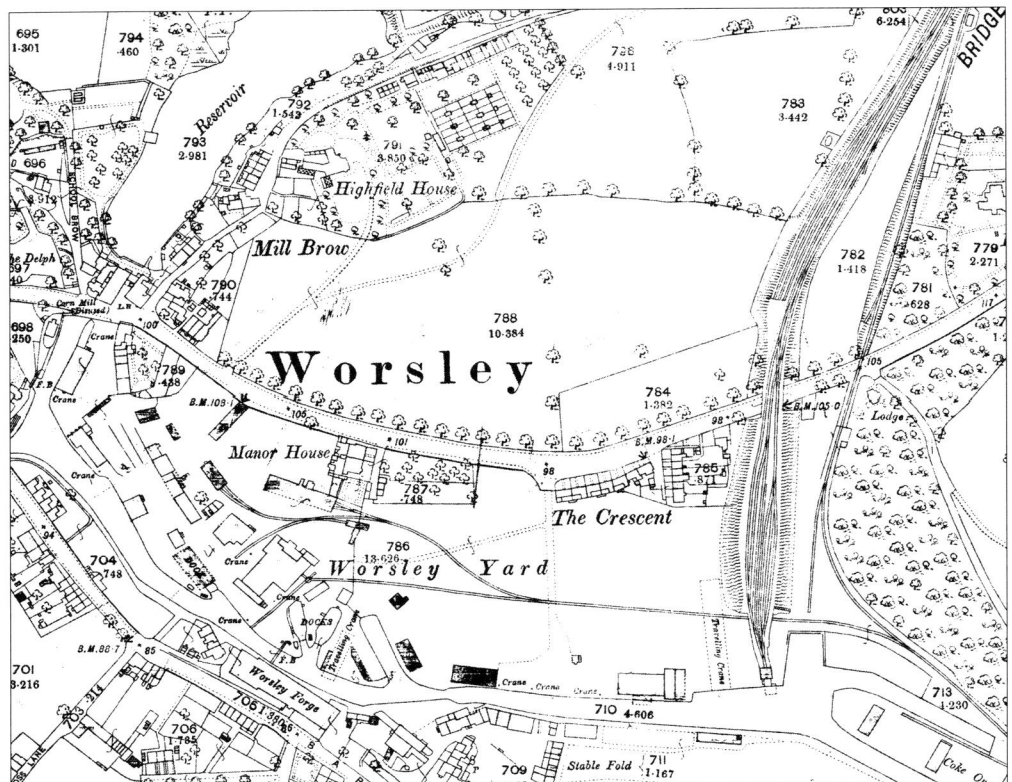

The 25-inch OS plan gives an indication of the hive of activity that was Worsley Yard in
1893. Top left is the Delph, where the underground canals branched northwards. Nearly
400 workmen were on site – mechanics, boiler smiths, joiners, saw millers, wheelwrights,
boat builders, saddlers, sail makers, wagon builders, locksmiths, painters and plumbers.
Note the canalside coal drops and coke ovens.

Carbon magazine in June 1935:

> Worsley Yard in those days was a bee hive of industry, employing upwards
> of 400 workmen and comprising in its various crafts mechanics, boiler
> smiths, joiners, saw millers, wheelwrights, boat builders, saddlers, sail
> makers, wagon builders, locksmiths, painters, plumbers and last but not
> least blacksmiths.
>
> At that time the smithy had a splendid set of fine built men ranging
> from the smallest of about 12 stone, up to the big forge man Mr James
> Tonge who scaled over 24 stone, and who at that period forged all their
> own large iron slabs.
>
> Another curiosity was William Peto Johnstone, commonly known as
> 'Scotch Bill' who made all the hurdles surrounding the shrubbery in front

A good clear early view of Worsley Yard from the south-west canal side. A covered dry dock is to the left and beyond the dry dock can be seen large stacks of pit props.

of [the] Bridgewater Offices, and hereby hangs a tale, which the writer can verify as true.

When Old Bill proceeded to erect the hurdles he and his striker went to Mr Joseph Gregory's at the 'Bulls Head' Hotel for their dinner. Old Bill ordered two pints of ale. Joe Gregory asked who the two pints were for. Old Bill said 'Me and him', pointing to his striker. The landlord told Old Bill he could have two, but the striker could not have any, as they did not serve lads at the 'Bull'. The striker was very young looking, but he was a married man 27 years old with two children.

The Yard was not closed for a holiday on Jubilee day, but an order was issued at 9.00 a.m. that all the workmen could have a 'go-as-you-please' day, but were not expected to leave the premises. A most enjoyable day was spent in games about the Yard. At that time the late Earl and Lady Ellesmere were constant visitors to the works when in residence at the Hall.

Most of the men working at Worsley would naturally be moving the 2 miles up the road to Walkden Yard on its completion in 1898.

An advertisement for Hudswell Clarke from *The Engineer* of 1896. The company was to supply locomotive 0-4-0 ST *Westwood* to Moston Colliery in 1913, the two Austerity design 0-6-0 ST locomotives *Allen* and *Charles* in 1944 for general use on the Central Railways and the 0-4-0 ST *Carr* in 1948 which worked most of its life at Astley Green Colliery.

An advertisement for Hunslet from *The Engineer* of 1896. Among others supplied to the Lancashire Coalfield, Hunslet supplied the following locomotives; the two 0-6-0 Side Tanks *Bridgewater* and *Joseph* to the Central Railways in 1924 and the 0-6-0 ST *Stanley* to the Central Railways in 1945. Hunslet were later to become major manufacturers of underground diesel locomotives.

The New Walkden Yard, 1897

After detailed planning lasting a few years, work on the new Walkden Yard had begun in late 1897. The *Farnworth Journal* announced that progress on the project was advanced on 7 May 1898 with the headline 'New Yard For The Bridgewater Trustees'. It commented that the new yard would be situated near the Lancashire & Yorkshire and London North Western lines and would be used by a large staff of joiners, carpenters, stonemasons, bricklayers, mechanics and engineers. Specialist workshops were being constructed along with an engine depot for the Trust's own locomotives. A superintendent's residence would be constructed.

Also on site at Walkden were a wheelwright's shop, a wagon machine shop and a sleeper creosoting plant. The transfer of activities from Worsley to Walkden was mostly completed by summer 1900. The local press announced on 14 July 1900 that 'The Bridgewater Trust having decided to concentrate their works at Walkden where an extensive set of workshops have been erected, the extensive yard at Worsley is now done away with. Instead of the busy hive of industry presented for many years, the forge fires are now out'.

In the meantime, in 1903 the terms of the 3rd Duke of Bridgewater's will expired. The remaining Bridgewater interests, namely farms, cottages, coal mines, brickworks, lime kilns, minerals, railways and residences came under the ownership of the Earl of Ellesmere.

Walkden Yard was to become the main locomotive shed for the Bridgewater fleet, with other sheds situated at Bridgewater Colliery (Sandhole) and Mosley Common Colliery. The old Worsley Yard had been cleared by around 1905, the chimney base being retained to this day as part of a memorial to the Canal Duke.

An interesting occurrence took place at Walkden Yard in 1907. On a Saturday afternoon on 22 June some men were breaking up scrap iron at the yard and decided to use mining explosives (Roburite) rather than sledgehammers, probably obtained from the adjacent Ellesmere Colliery. It appears a misjudgement (to put it mildly!) was made in the amount of Roburite required. The result was a large crater; several large pieces of iron were blown 'a considerable distance' in the air. One piece fell among a group of Lord Ellesmere's clerks on the bowling green next to the Bridgewater Offices; another piece landed half a mile away; another smashed a cottage window; and another, weighing 16 lbs, fell in Stanley Road (over 500 yards to the south-east) hitting a lamp post. Several persons reported having narrow escapes.

The original Earl of Ellesmere's Fire Brigade appliance was a Merryweather type, seen here outside Bridgewater Collieries Offices in about 1900. It was later based at Walkden Yard and was replaced shortly after 1924. (Bessie Hilton)

Ellesmere Colliery is shown looking south in July 1905 by Eccles photographer Harry Grundy. Inclusion of the area to the right would have given us the earliest view of Walkden Yard. Note the wooden pit headgears and the 1866 vertical Naysmyth steam winder engine house serving both shafts at once, with a single cage in each shaft. Note also the Bridgewater Collieries wagons and, in the distance, the Bridgewater Collieries Offices, with the Lady Bourke clock atop.

		£	s	d
1908	New Fence Wall ~ Walkden Yard.	194	17	0
"	Horizontal Slot Drilling & Keyway Cutting Machine for Mechanic's Shop W'den Yard	119	0	0
"	One 10 ton Steam Hammer for Blacksmith Shop. Walkden Yard.	168	0	0
"	One Radial Drilling Machine for Mechanic's Shop. W'den Yard.	160	0	0
"	One Patent Planing Machine for Mechanic's Shop W'den Yard.	155	0	0
"	New Electric Travelling Crane for Blacksmith's Shop.	85	0	0
1907	3 ton Motor Lurry from Commercial Cars Ltd. to Walkden Yard.	795	0	0
1909	1 ton Motor Lurry from Commercial Cars Ltd. for Walkden Yard	439	4	0
1908	New Store Room Walkden Yard.	330	3	0
1909	New Crane in Mechanic's Shop.	68	9	0
1911	460 New Pit Tubs made in W'den Yard for Ashtons Field Pit to	1150	0	0
"	50 New Pit Tubs (7 cwts) from W'den Yard for Brackley Pit	125	0	0
"	Roof over Boilers at Bridgewater No 3 pit & Booth & Sons.	167	9	11
"	Induced Draught Plant for Mosley Common No 1 & 2 supplied by Messrs Matthews & Yates.	305	0	0
"	300 Steel Tubs 20 cft Capacity for Mosley Common No 4 pit	1162	10	0
"	46 Well Bottomed Do Cft capacity Pit Tubs for Mosley Common No 1 & 2 pits	178	5	0
"	50 New Tubs from W'den Yd for Wharton Pit	125	0	0
"	2 Turntables for Miles Platting Wharf	149	0	0
	Labour on above	32	10	0
"	One Wagon Weighbridge for Broadheath Wharf	55	12	6
"	New Rotor Motor for Mechanic's Shop.	105	0	0
"	1- 5 B.H.P. 3 Phase Standard Motor for Smithy, Walkden Yard.	24	5	0

It's spend, spend, spend as Walkden Yard is further kitted out. In the Heavy Expenditure records from 1908 to 1911, we see purchases of drills, electric motors, a planer, a steam hammer, electric overhead crane, a 1 and a 3 ton 'Motor Lurry' and a slotting machine. The record also shows that Walkden Yard was manufacturing pit tubs – 460 were made for Ashtons Field Colliery in 1911. (Lancashire Archives, NCB Bw 4 Heavy Expenditure 1907–16)

By approximately 1908 a new mineral line had been laid out to link Brackley Colliery and nearby Wharton Hall Colliery to its south, west of Little Hulton, to the Bridgewater system. The vast underground canal system had by now reached as far as Brackley Colliery, slightly beyond Little Hulton to the west.

On the death of the 3rd Earl of Ellesmere on 13 July 1914, his son succeeded him as the 4th earl and inherited all his estates and financial interests. A detailed valuation of the company's assets (now held at the Lancashire Record Office) after the death of the earl in 1914 showed the following collieries in operation: Ashton's Field, Ellesmere, Linnyshaw, Brackley, Wharton Hall, Bridgewater (Sandhole) and Mosley Common.

Hundreds of other unnamed shafts were still accessing the vast multi-horizon mining complex, which at its peak was probably the most extensive coal mining system Britain has ever seen.

The company owned 1,742 wagons plus 142 for internal use. On the canal the company worked a steam tug, forty-three lighters of 40–58 tons capacity, two Liverpool flats of 64 tons capacity, two 'bastard' boats of 42–45 tons capacity, twenty-three box boats accommodating ten coal boxes of 2 tons capacity, fifty-six 16-ton 'B' boats and twenty 10-ton 'M' boats. The company also owned its own brickworks at Ellenbrook near Mosley Common.

The Bridgewater Collieries 'workmen's special', hauled by loco *Francis* (Manning Wardle 311/1870), is seen possibly at Ashtons Field Colliery in about 1900, miners on board. This ran all the way from Worsley Yard to Mosley Common Colliery, then up to Sandhole Colliery, Ashtons Field Colliery (allowing reversal), and Walkden Yard. The arrival of trams ended its working life. The loco was sold to T. W. Ward of Sheffield in 1920 for £200, and then to Charles Walmesley & Co. Ltd of Bury in June 1926.

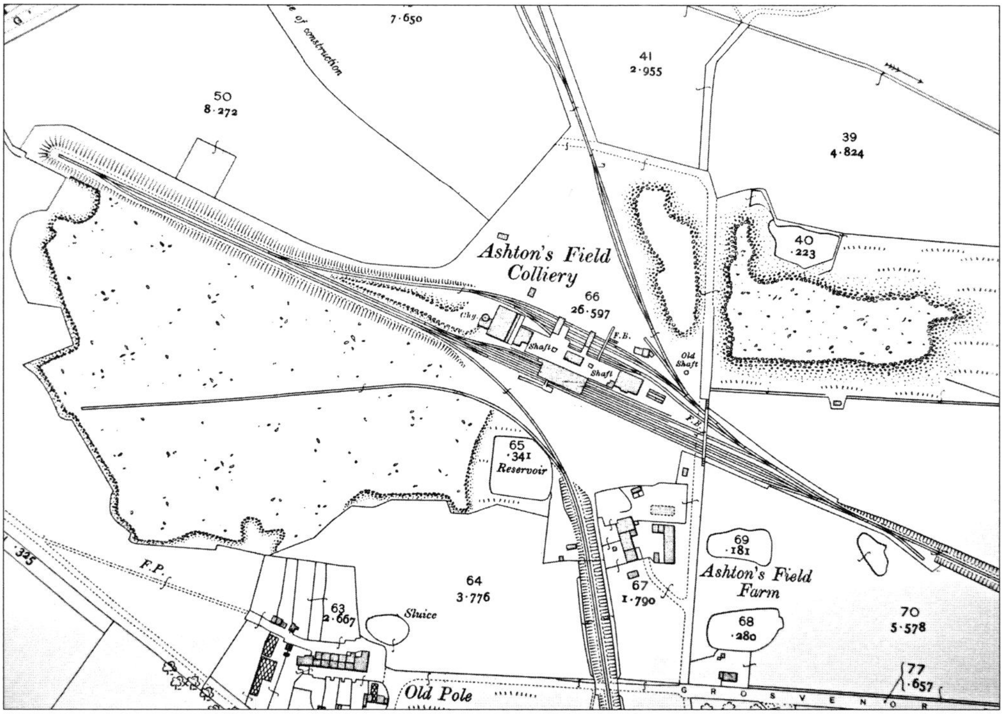

The reversal point for the workmen's special train at the Ashtons Field Colliery site in 1907. Waste tips grow to the west (only recently reclaimed) and to the right is the line to the Kearsley Branch of the L&YR and Linnyshaw Moss. At the bottom can be seen the level crossing over Grosvenor Road, north-west of Walkden.

In the earliest photograph we have of Walkden Yard workmen, we see the forge men in 1908, probably ex Worsley Yard employees who arrived in 1900. These were skilled men, creating and shaping metal forms for hundreds of different uses above and below ground at the collieries. The men would be members of the Associated Blacksmiths' Society (union).

Fred Millington, the first Master of the Yard at Walkden Yard, stands central in his Earl of Ellesmere's Fire Brigade Chiefs peaked cap in about 1910. His company-provided home, Tynesbank House, is behind at the entrance to the Yard. Note the old horse-drawn Merryweather type fire engine, replaced after 1924. (Bessie Hilton)

Bridgewater Collieries' records (NCB Bw) at the Lancashire Record Office show investment taking place post-First World War during the period of part-nationalisation the company had 'suffered' (until 1921). Mining companies were restricted in the amounts of profit they made during the war and its aftermath, so it appears Bridgewater Collieries decided it was better to reinvest profits into the company infrastructure rather than paying their excess profits to the Treasury!:

> New log saw mill in course of erection December 1920, brick with part glazed and part corrugated asbestos roof.

The earlier boatshed workshop site to the north of the new Walkden Yard gradually became a coal landsale yard (surviving as such until around the mid-1970s).

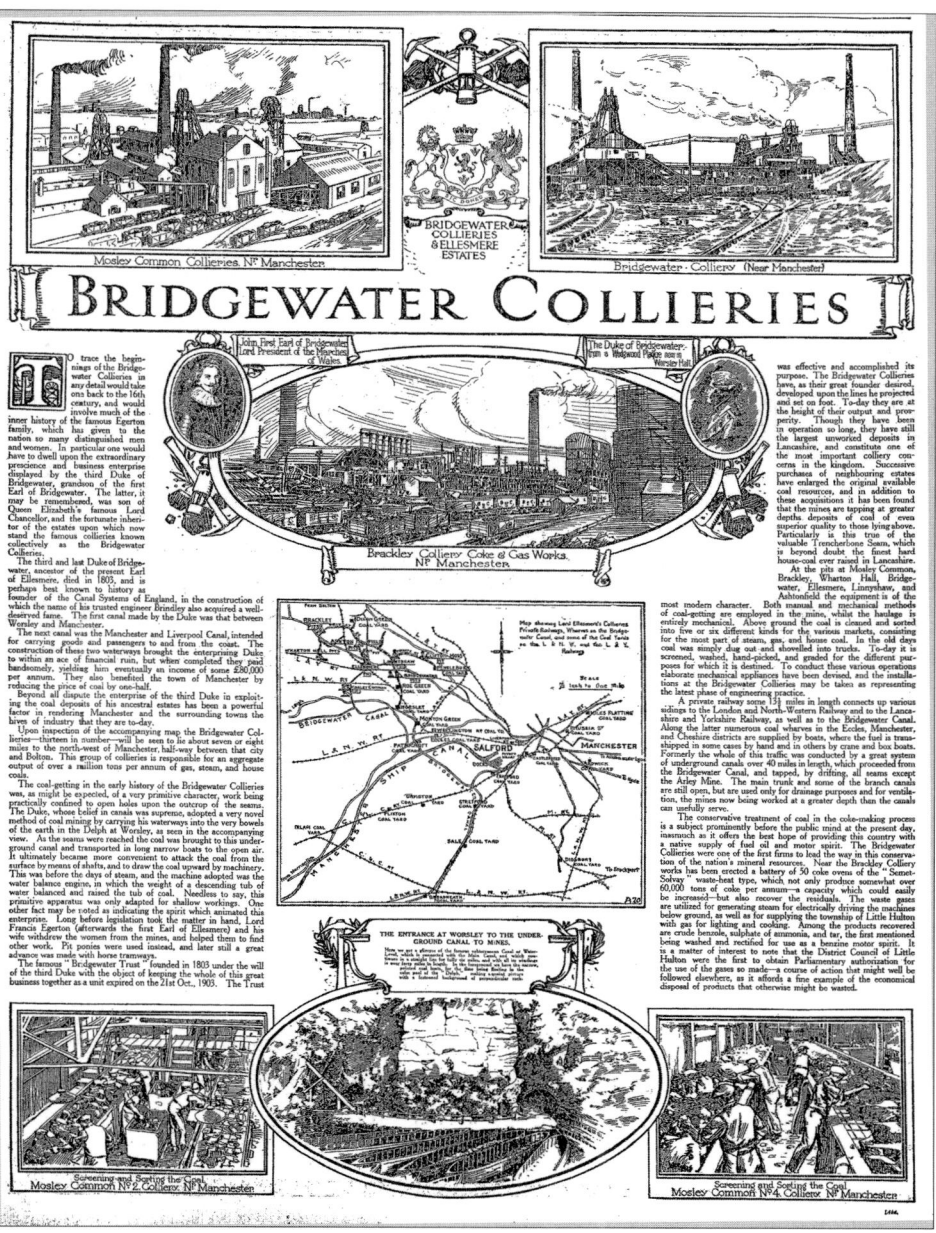

1913 saw the peak of British coal production, at over 287 million tons. The Bridgewater Collieries mining complex was one of the largest in Britain, a feature on the company being published in *The Times* on 1 December 1913. The artist has depicted Mosley Common Colliery top left, and Bridgewater (Sandhole) Colliery top right. Top centre is an illustration of the Brackley Colliery coke and gas works, which produced oil, petrol, ammonium sulphate and tar. The (then) 13½ miles of private mineral railway is shown on the plan lower centre. At the base are illustrations, left and right, of Mosley Common Colliery screen hands and, centre, the entrance to the underground canals at Worsley Delph.

In March 1915 it had become clear there was an acute shortage of munitions due to the large number of male workers enlisting in the forces. The ensuing outcry became known as the 'shell scandal'. Walkden Yard took on munitions work, producing shell cases. Shifts were often longer than normal on munitions work, up to twelve hours.

The Bleak 1920s: Company Reorganisation

The Coal Mines (Emergency) Act of 1920 controlled the profits and wages systems in the industry. The Mining Industry Act 1920 was passed 'To provide for the better administration of mines, and to regulate the coal industry, and for other purposes connected with the mining industry, and the persons employed therein'.

These additional post-war controls on the industry understandably worried the coal owners and colliery proprietors, who thought that they were in danger of gradually losing their grip on their highly-lucrative livelihood; it was time for change at Walkden. Investment in the existing infrastructure carried on as normal in the meantime with the major addition of the saw mill, constructed from December 1920 onwards, which was fully equipped by December 1921 to the tune of over £6,600.

LOCOMOTIVES.

ESTIMATED VALUES AS AT 31st DECEMBER, 1920.

	£.	s.	d.
Francis	4,000.	0.	0
Stanley	700.	0.	0
Grand Duchess	700.	0.	0
Sultan	1,000.	0.	0
Alert	1,000.	0.	0
Atlas	1,000.	0.	0
Gower	1,000.	0.	0
Wardley	1,000.	0.	0
Madge	2,000.	0.	0
Violet	2,000.	0.	0
Brackley	2,500.	0.	0
Ellesmere	2,500.	0.	0
Katharine	3,000.	0.	0
Carried to Summary	£22,400.	0.	0.

Note.- The amount of spares to above
at Dec. 31st, 1920 may be taken
as "Nil".

Enamel sign made from the 1920s onwards and seen around the perimeters of the Bridgewater private estates. These signs would certainly be memorable to many a former poacher and are still to be found today.

1921 Bridgewater Collieries and Wharves

In 1921 the 4th Earl of Ellesmere created the Bridgewater Collieries and Bridgewater Wharves companies. By doing so these interests were isolated from his main estate, giving protection against potential death duties, taxation and possible fragmentation of the estate due to future nationalisation of the mining industry. The mines had in effect been nationalised during the First World War until they were returned to the mine owners in 1921, hence the fear and suspicion.

The older collieries were coming to the end of their lives. On 31 March 1921 Linnyshaw Colliery closed; Bridgewater Colliery (also known as Sandhole Colliery, and to the south of Linnyshaw) then accessed its remaining coal reserves. Pumping water from the shafts at Linnyshaw to ease the burden on Sandhole continued until 1936. Ellesmere Colliery, adjacent to Walkden Yard, also ceased production in 1921 (being retained as a pumping station until 1968). The retention of sites for pumping and the ongoing cost of operating them gives an idea of the immense volumes of water to be dealt with in the vast Bridgewater mining complex.

In 1923 Bridgewater Collieries Ltd and Bridgewater Wharves Ltd were sold with the estate to the new Bridgewater Estates Ltd, becoming subsidiaries (later to be given up).

Ashton's Field Colliery closed on 22 May 1925; it reopened in 1926 but was abandoned in November 1929. In September 1925 the company purchased Newtown Colliery from Clifton & Kersley Coal Co. Ltd, situated north of Swinton, allowing Bridgewater Colliery (Sandhole) to access previous boundary or barrier coal in various seams between the pits.

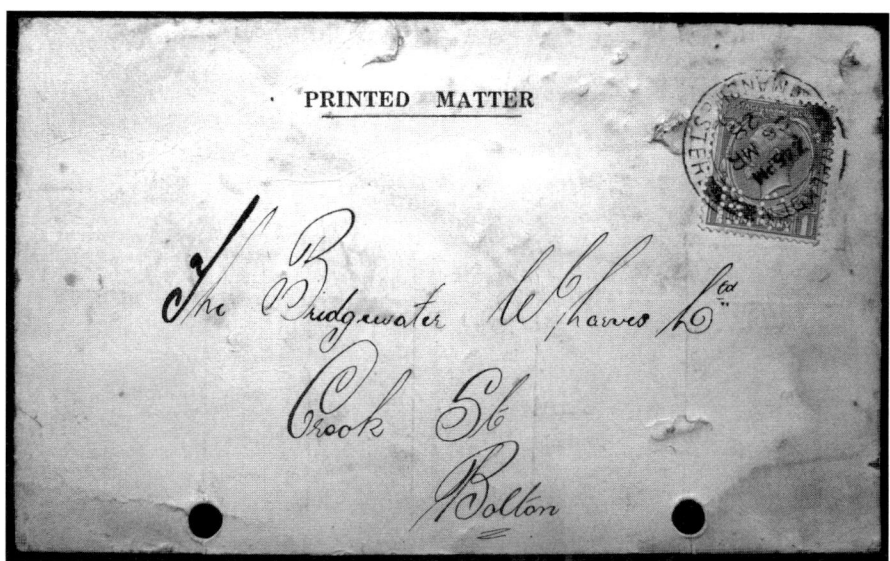

A year before the creation of Manchester Collieries, this coal wagon invoice card accompanied a delivery by Bridgewater Collieries to the Bridgewater Wharves coal depot at Crook Street, Bolton. (Steve Leyland)

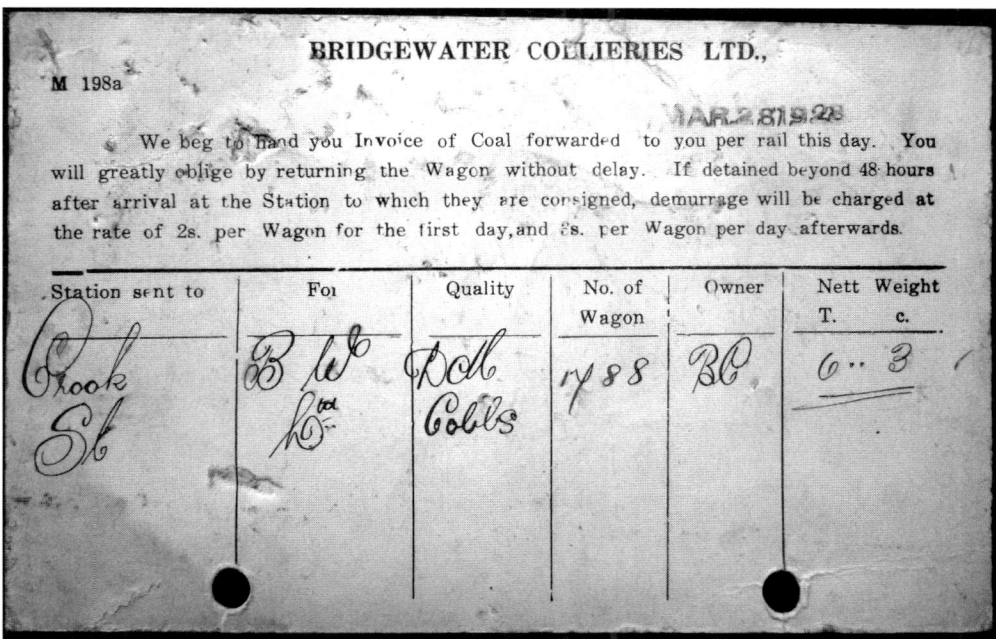

The reverse of the 1928 Bridgewater Collieries coal wagon card shows 6 tons and 3 hundredweights (20 hundredweight = 1 ton) of coal was delivered to Bolton Crook St Wharf (depot).

BOY KILLED BY RAILWAY TRUCK.

Coroner's Distress.

Mr. J. Fearnley, the Bolton Borough Coroner, conducted an inquest yesterday on a boy whom he knew personally, and it was with great emotion that he said at the close of the inquest: "He was one of my own Wolf Cubs. He was a very nice lad—I can't say anything more." Mr. Fearnley is connected with the Boy Scout movement in the Walkden district, where he lives.

The boy was Leslie Davies, aged 12, of 47. Gower Street. Little Hulton. It was stated that while he was playing near a private railway line owned by the Bridgewater Colliery Company, Limited, at Little Hulton he and another boy ran to the fence to watch a train pass. It was nearly past when Davies cried: "Just watch me have a ride on those buffers!" He ran down the embankment and caught the buffer of the last wagon. Just then the train stopped and jerked him off, and the wagon, rebounding, ran over him.

A verdict of accidental death was returned.

A tragic accident. The extent of the mineral line system meant that children had many opportunities to wander on and play or dare each other.

The Mining Industry Act 1926 proposed to stem the post-war decline in coal mining (the strikes of 1921 and 1926 provoking a number of colliery closures) and encourage independent companies to merge in order to modernise and better survive the desperate economic conditions. Problems were not confined to the actual working of the collieries: a shortage of main-line wagons even leading to temporary colliery closures in the coalfield at Leigh, Haydock and Over Hulton in January 1927. In the meantime Wharton Hall Colliery, west of Little Hulton and a few hundred yards south of Brackley Colliery (and today's A6), ceased production on 23 December 1927, becoming yet another pumping station for the mining complex and relieving the burden on Brackley Colliery.

The Lancashire Electric Power Company: A Major Customer

The LEP was one of the largest private electricity companies in the UK, established by Act of Parliament in 1900. Radcliffe Power Station was the first of this enterprising company's sites, opened by the Earl of Derby on 9 October 1905. In a parallel with modern times Walkden Yard 'switched' to having its power supplied by the LEP in 1912. The company opened another power station at Padiham near Burnley in 1926. Manchester Collieries, incorporating Bridgewater Collieries, was formed in March 1929. The new company was aware of the huge benefits of regular power-station coal sales to their own power supplier. One of their new member collieries, Outwood, was only ¼ mile down the line from Radcliffe Power Station, but was to cease production by February 1931.

Continuity of power-station sales after the closure of Outwood had been maintained shortly after Manchester Collieries was formed, as the LEP's third site, east of Farnworth at Kearsley, had opened in November 1929. Kearsley was linked to the mineral lines via the Kearsley Branch of the LMSR (formerly L&YR) north of Linnyshaw Sidings. Initially, raw or, what is termed in mining, 'Run of Mine' coal (unscreened) was sent to Kearsley from Astley Green Colliery, Mosley Common Colliery and Brackley Colliery, the last leg carried out by the LMS. On 31 March 1937 a coal-blending plant began work at the site of the former Ashton's Field Colliery (ceased production November 1929), holding and processing coal from the three pits with Linnyshaw Sidings close by. Lancashire Electric Power was nationalised in 1948, forming part of the North Western Electricity Board. Radcliffe Power Station closed in 1959, Kearsley Power Station in 1981.

Manchester Collieries Ltd: The 1930s Heyday of Walkden Yard

Chanters Colliery, Atherton, was for generations operated by Fletcher Burrows & Co. Ltd. In March 1929, they amalgamated with five other coal companies to form Manchester Collieries. Here, in the October 1929 edition of *Carbon* magazine, some of the member companies' wagons are seen alongside Chanters coal washery – Atherton Collieries, the new Manchester Collieries wagon, Bridgewater Collieries, Astley & Tyldesley Collieries and Andrew Knowles & Sons Ltd.

March 1929 saw the end of an era for Bridgewater Collieries Ltd with the formation of Manchester Collieries Ltd. This amalgamation had been proposed as far back as 1923 by Robert Burrows of Fletcher Burrows & Co. Ltd, Atherton, and separately by Joseph Ramsden, chairman of Bridgewater Estates, no doubt reeling after the experience of First World War 'nationalisation', which only ended in 1921.

The wealthy and politically-ambitious Robert Burrows, later *Sir* Robert, was very well connected, even moving in royal circles. No doubt he would have had fore knowledge of the proposed Mining Industry Act 1926 and its intention to promote nationalisation and amalgamations within the industry.

As late as 20 August 1928 the *Manchester Guardian* reported that only three companies were initially intended to merge: Astley & Tyldesley Collieries Ltd, Andrew Knowles & Sons Ltd and the Bridgewater Collieries. The *Manchester Guardian* announced on 27 February 1929 that six companies, along with

Manchester Collieries' engineering managers pose for the *Carbon* magazine photographer at the Walkden offices in July 1932. From left to right, back row: Fred Hilton, Walkden Yard manager; R. Yeaman, Central District; A. Morris, Eastern District; W. Kearsley, Electrical Engineering. Front row: S. Greenhalgh, Chief Engineer; F. Willink, Engineering Executive; Tom Fox, Western District (Gin Pit Workshops).

Manchester Collieries' sales office at 28 Cross Street, Manchester, seen in *Carbon* magazine in July 1932. In the window are large scale models of the company's wagons along with the humorous mining related illustrations W. Heath Robinson produced for an Atherton Collieries calendar in 1922.

four of their subsidiaries, were to merge with a total capital of £7,000,000. Mr Justice MacKinnon of the Railway and Canal Commission (who sanctioned the merger) questioned quite rightly why no less than twenty-three directors of the new company were necessary!

To the west of the coalfield in early 1930 another substantial merger took place, Wigan Coal Corporation being formed through the merger of the Pearson & Knowles Coal and Iron Co. Ltd and Wigan Coal & Iron Co. Ltd, comprising twelve collieries and 11,760 men.

The Manchester Collieries merger was to reaffirm the important role Walkden Yard had played in locomotive and colliery engineering maintenance. The six constituent colliery companies and their subsidiaries (with two added later in 1934 and 1935) had amalgamated to form one of the largest colliery concerns ever seen in Britain. The intention was to improve marketing and sales, to standardise working methods and technology across the coalfield, to centralise workshops provision and power distribution. This was a timely move as the forthcoming Coal Mines Act 1930 was to further increase the pressure for colliery amalgamations.

Manchester Collieries Ltd initially comprised Astley & Tyldesley Collieries Ltd of Tyldesley; Bridgewater Collieries & Wharves Ltd, Walkden; Clifton & Kersley Coal Co. Ltd of Clifton near Swinton; Fletcher Burrows & Co. Ltd of Atherton; Andrew Knowles & Sons Ltd of Pendlebury near Swinton and John Speakman & Sons of Bedford, Leigh.

Until 1929 most coal had been won traditionally by colliers with explosives, picks (occasionally pneumatic), shovels and tub haulage. Manchester Collieries embarked upon a major programme of mechanisation with machine-cut coal increasing in percentage from 17 per cent to 98 per cent in sixteen years. The haulage of coal below ground was to be rapidly mechanised, with pit ponies at a few of the older collieries being phased out by 1932.

In 1933 the company was divided into five groups totalling sixty-one shafts with coal winding taking place at twenty-seven of these. Over 15,000 men were employed by the company working ten different seams of coal. Around 250 miles of underground roadways were in use, along with 257 coal cutters and 600 light mechanical picks and drills. ninety conveyors were in use at the coal face, over 700 electric and compressed-air haulage engines, 8,000 miners' oil lamps and 11,000 electric lamps (cap lamps and 'bottle' lamps) maintained. A central laboratory at Walkden constantly analysed and publicised the results of coal quality tests from the ten seams being mined, a vital requirement of customers at this time.

Collieries served by Walkden Yard now included Agecroft (Salford), Ashton Field (Walkden/Farnworth) Astley Green (south of Tyldesley), Bedford (east of Leigh), Brackley (Little Hulton/Farnworth), Bradford (east Manchester), Chanters (east Atherton), Ellesmere (Walkden), Gibfield (west Atherton), Gin Pit (south Tyldesley), Howe Bridge (west Atherton), Ladyshore (east Bolton), Mosley Common (east Tyldesley), Newtown (north Swinton), Nook (south Tyldesley), Outwood (Radcliffe), Pendleton (west Manchester), St Georges (south Tyldesley), Wharton Hall (north-east Tyldesley) and Wheatsheaf (north Swinton). Also served were Worsley Boatyard, Poynton Yard and Robin Hood coal washery (north Swinton).

It was stated the company operated ninety-eight miles of private railways with thirty-eight locomotives and 10,000 wagons. About 280 canal barges were based at the five canal basins belonging to the company, as well as eighty motor vehicles and 375 horse-drawn wagons. By now Walkden Yard consisted of a heavy machine shop, light machine shop, compressed-air hammer and forge shop, joiner's shop, electrician's shop, paint shop, springsmith's shop, sawmills, tinsmith's shop, wagon repair shops, wagon machine shop and coal-analysis laboratories.

The emergence of Manchester Collieries Ltd brought with it a feverish period of investment in technology standardisation and innovation. Progress was highlighted in *Carbon* magazine, formerly the journal of the Atherton

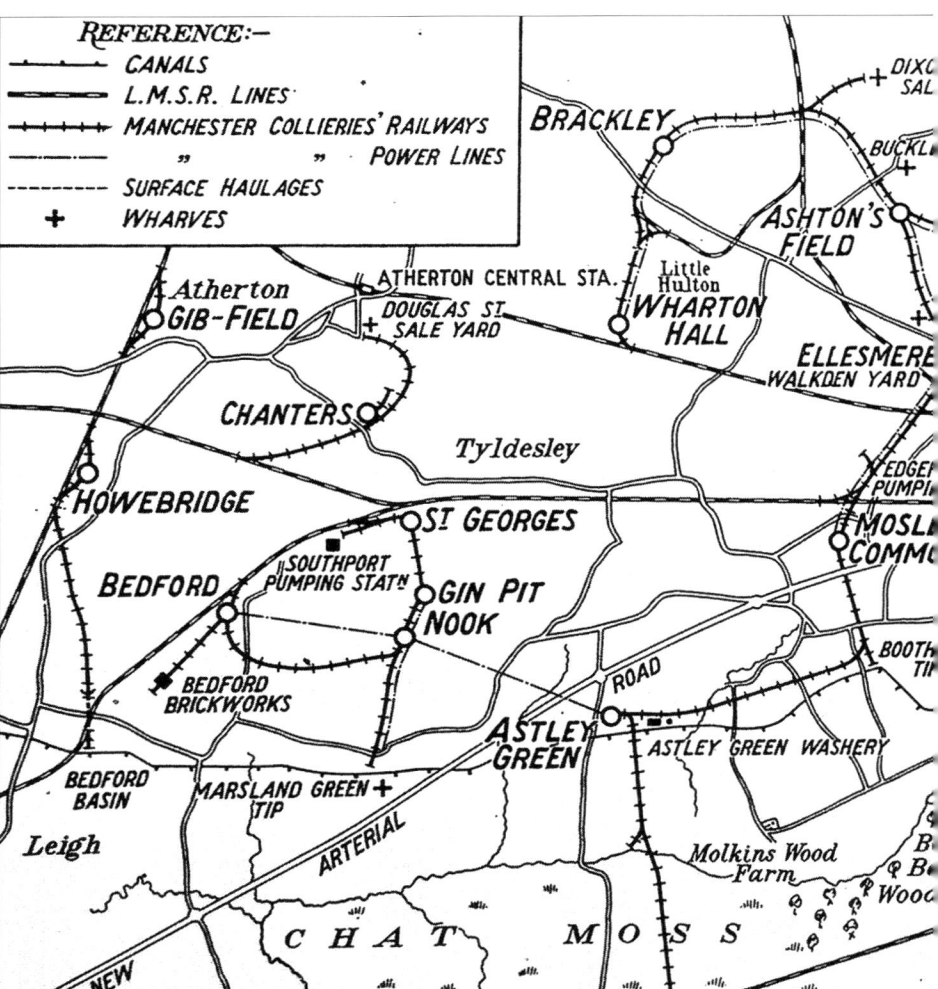

The full extent of the mining and rail network in Manchester Collieries days can be seen in this 1933 map. This was included in a feature on Mosley Common Colliery in *Colliery Engineering*.

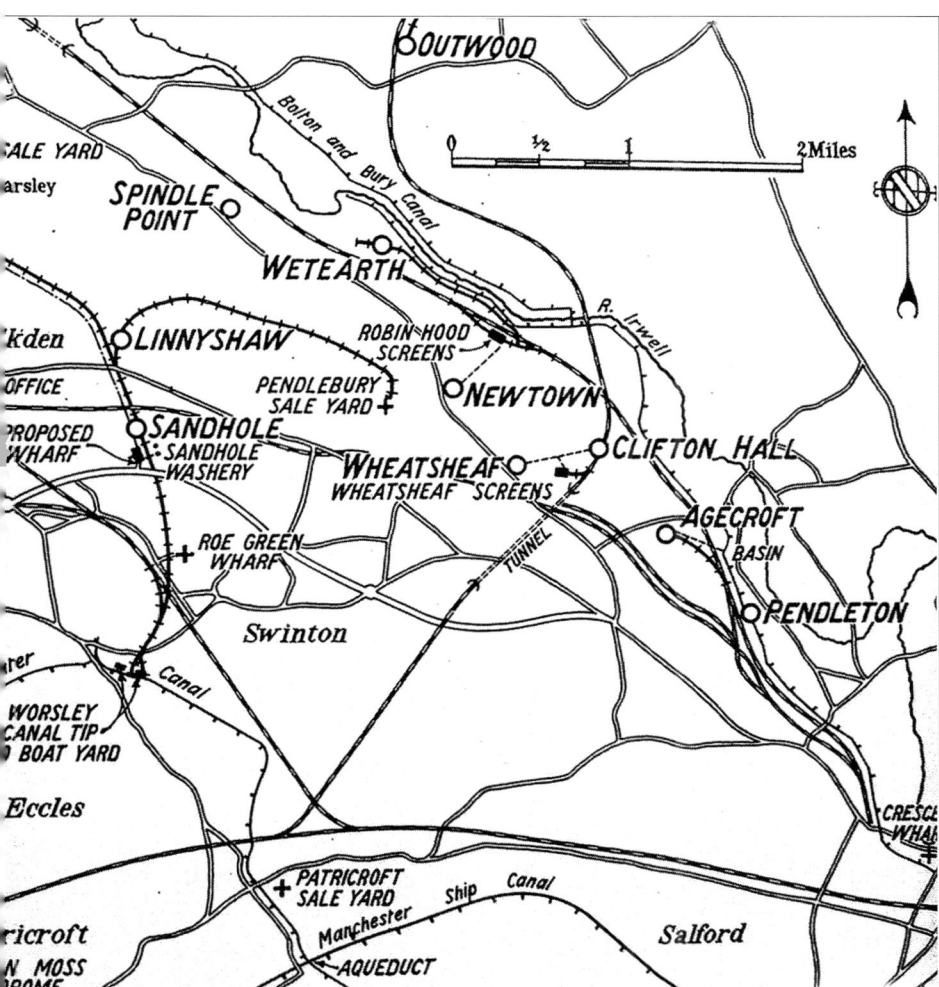

If its from Manchester Collieries its right

Producers and
Distributors

of all
classes of

HOUSE
GAS

MANCHESTER
COLLIERIES LTD.

Telephone: Blackfriars, 8633.
Telegrams: Fuel, Manchester.

Head Sales Office:
Arkwright House,
Parsonage.
Manchester.

and

STEAM
FUELS

QUALITY and SERVICE

Manchester Collieries' advertisement from their magazine *Carbon*, 1932

Collieries operated by Fletcher Burrows & Co. Ltd. *Carbon* magazine is a source rarely found in industrial history research today, but is one combining the social history of the workforce and the progress of the business.

Arthur McCullough looked back in 1989 after the yard had closed (in August 1986) to his employment at Walkden Yard from 1929 to 1954. Arthur perhaps did not enjoy his time at Walkden and does not mince words! Interestingly he gave the drawing office as much praise as the skilled engineering workforce. Hours of work were seven in the morning till half past four in the afternoon, with half an hour for lunch. Saturday mornings were worked from seven until twelve. Wages were dictated by management, they were kept very low, forcing men to rely on overtime. He recalled overtime was only granted on condition of good behaviour and attitude. He states that it was not until 1932 that a canteen and decent toilets were provided. He remembers quite a number of men being sacked. It was done quite discreetly, no warnings, just a note in your pay packet on a Friday to say your services were no longer required. Arthur adds, 'Management was completely unapproachable and riddled with old Victorian snobbery. Promotion was a mystery'. The fact that Arthur left Walkden Yard in 1954 to teach carpentry in prisons appears to support his experiences, although equally my conversations with those related to former managers at the yard and with some employees give the completely opposite picture!

The long established Charles Roberts wagon works of Wakefield supplied over 300 wooden framed 12-ton wagons to Walkden Yard under Manchester Collieries in the 1930s. In 1933 the company owned approximately 10,000 wagons, supplied by at least four companies.

A workforce 'community' of a sorts was gradually built up over the years at Walkden Yard, occasionally offering the chance to display team spirit. *Carbon* magazine mentioned in April 1932 that an ambulance team had been organised, coming second to the virtually-unbeatable Mosley Common Colliery No. 4 Pit team.

At this time the Ashton's Field (Colliery) pit tub fabrication shop a mile to the north of Walkden had been busy manufacturing a standard tub for the company, utilising its 700-ton hydraulic press. Variations in dimensions and capacity at the many colliery companies along with varying surface tippler design had hampered standardisation. Eventually a design was completed and went into operation at Ashton's Field. By April 1932 over 4,000 tubs had been built at a rate of 150 a week. Hooks and drawgear had been constructed of a new steel, titled 'Chromador', with tests also being carried out at Manchester College of Technology. The use of Chromador on the Sydney Harbour Bridge, no less, had inspired engineers at Manchester Collieries to try it out.

By December 1932 *Carbon* was announcing that the main-line wagon repair shop at Walkden had been extended. Facilities for spray painting of wagons had arrived 'showing considerable savings compared with the older hand painting methods'. In September 1933 it was announced that the wagon fleet was gradually being repainted, with over a thousand having been completed over the previous six months. The company's horse-drawn carts and lorries were also being dealt with in the company's colours of 'red flame and black'.

The Lancashire collieries were by now divided into three groups; the Central Group comprised the former Bridgewater Collieries and mineral rail network based around Walkden, the network to be termed Central Group Railways (later, in post-1947 NCB days, becoming just Central Railways).

A Manchester Collieries four-wheel, horse-drawn 2-ton coal delivery wagon from *Carbon* magazine of September 1934. The upgrade to pneumatic tyres was in full flow at this time. Walkden Yard built, painted and repaired these wagons.

A Manchester Collieries two-wheel lump coal wagon with tipping capability, seen in *Carbon* magazine in September 1934.

W. Ramsden & Sons' ailing Shakerley Collieries Ltd, Tyldesley, joined Manchester Collieries in 1935. The company was added to its portfolio to offset the restrictive company coal quota system in place under the Coal Mines Act, 1930. A four-plank 10-ton coal wagon is seen after refurbishment at Tyldesley Goods Yard, Well St, in about 1911.

Under the Coal Mines Act 1930 a standard production quota had been specified for the size and output of a colliery. The Act brought in quotas to prevent overproduction of coal and a statutory system of minimum prices was put in place to prevent damaging, aggressive competition. Manchester Collieries cleverly added 500,000 tons to their total coal production allocation by purchasing the ailing Shakerley Collieries, Tyldesley, owned by William Ramsden in 1935, whose Wellington Pit closed in the same year with its other mine, Nelson Pit, closing three years later. Bradford Colliery, east of Manchester near Clayton, owned by a subsidiary of Fine Cotton Spinners was acquired in 1935, not this time purely to boost quotas as it had enormous coal reserves and would still be in production in 1968.

The Lancashire and Cheshire District (Coal Mines) Scheme of 1930 controlled coal output during this bleak period. Coal quality was now more paramount than ever in securing customers, especially those in other counties. Astley Green Colliery erected a new coal washery in 1931, its line to Boothsbank canal basin being completed in the same year. Joe Cunliffe, former Walkden Yard manager, recalled in 1993 that it was decided that at the opening of the line Manchester Collieries' directors and their wives should be taken by train to the main-line connection near to Astley Station, south of Astley Green Colliery. Some wagons were thoroughly cleaned and lined with white material, fitted with seats and with screw couplings borrowed from the LMS. Locomotive *Madge* (o-6-o STIC Naysmith Wilson, 782/1906) was newly ex-workshops and repainted so was selected to haul the train. Joe remembers being told to fill the tank and receiving an ear-wigging for climbing up to the top and risking scratching the paint, being told to get a ladder. At some point the train broke a tape stretched across the line to signify the formal opening. When the party returned they all went into the offices where, no doubt, refreshments were laid on.

Figures for year ending 31 March 1932 showed that over 200,000 tons of coal had been transported on the line from Astley Green. Locomotives based at Astley worked as far as Boothsbank where Walkden engines took over for the hard slog to the north. The new line was to be a cost-effective one with reduced yearly charges for the use of the LMSR sidings at Astley Moss, south of the colliery.

Luckily for the company certain nationally-renowned seams of coal were still available from their collieries, such as the Trencherbone, Rams and Arley. Lancashire Associated Collieries Ltd was formed in July 1935 to efficiently market coal produced from the whole coalfield including Cheshire, its slogan being 'Burn Lancashire Coal'. The company was to remain in operation until nationalisation of the industry in January 1947.

Centralised Wagon Repairs

Extensions at Walkden Yard to allow for centralised repairs were completed by December 1932; pit-tub manufacture and repair were to be concentrated at Ashton's Field. An idea of the volume of main-line wagon repairs at Walkden can be seen for the year ending March 1932 when 6,032 had passed through. The company stock also continued to increase in size, up to 8,619 in 1934 and 10,966 plus internal wagons by 1936.

By November 1936 it was stated by the Coal Mines Reorganisation Committee that the two giants, Manchester Collieries and Wigan Coal Corporation, were responsible for 43 per cent of the production of the Lancashire coalfield. The industry at this time sensed that long term the findings of the committee might eventually lead to public ownership of the coal industry. The Coal Mines Reorganisation Committee was abolished by the Coal Act 1938, which enabled the transfer of coal to public ownership. On 1 July 1942 the nation's coal reserves became vested in the Coal Commission, compensation amounting to £66,450,000 being paid to former royalty owners. The Coal Commission was not in place to 'mine' the coal, but rather to act as public landlords ensuring the resource was worked efficiently for the public good and especially for the health of the industry which was so vital at the time.

The George Booth
Reminiscences of Walkden Yard
in the 1930s

The dry official archives of activities at Walkden Yard over its life in the records of the Trustees of the Duke of Bridgewater, Bridgewater Collieries and Wharves, Manchester Collieries, Lancashire Associated Collieries, the National Coal Board and finally British Coal can never give us a feel for the reality of daily working life at the yard. The social history of the yard, the nitty-gritty of day-to-day life, is well and truly brought home through the well-composed, often humorous and now historically-important recollections of George Booth. George began his working life at Walkden Yard as an apprentice welder, aged fourteen, in early 1932.

In a series of booklets published by the Walkden Library Local History Group from 1988 George eventually supplied enough material for *Walkden Yard 1932–1939* (published in 1988), *Walkden Yard Yarns No. 1* (published in 1990), *Walkden Yard Yarns No. 2* (published around 1994), and *The Trustees Railways and After* (published around 1996).

George's recollections show the often tough and harsh side of work at a large and very busy colliery maintenance depot, the old-fashioned, hierarchical background of strict management, yet still allowing the emergence of memorable characters among the workforce.

The selections which follow highlight George's humorous style and his memory for a level of detail that locomotive enthusiasts will appreciate. His accounts of colliery and railway operations are of genuine historical importance, recording aspects that future historians will find nowhere else. It has been mentioned to me by relatives of former officials at Walkden Yard that George's account occasionally contains 'artistic licence', the odd inaccurate description of working practices or a false critical description of the character of members of staff, but during the course of researching this book I have been unable to meet anyone who worked in the yard in the period George writes about to verify this. Railway historians have been generous in correcting for me a few errors in George's account (in square brackets alongside), for which I am grateful. Writers

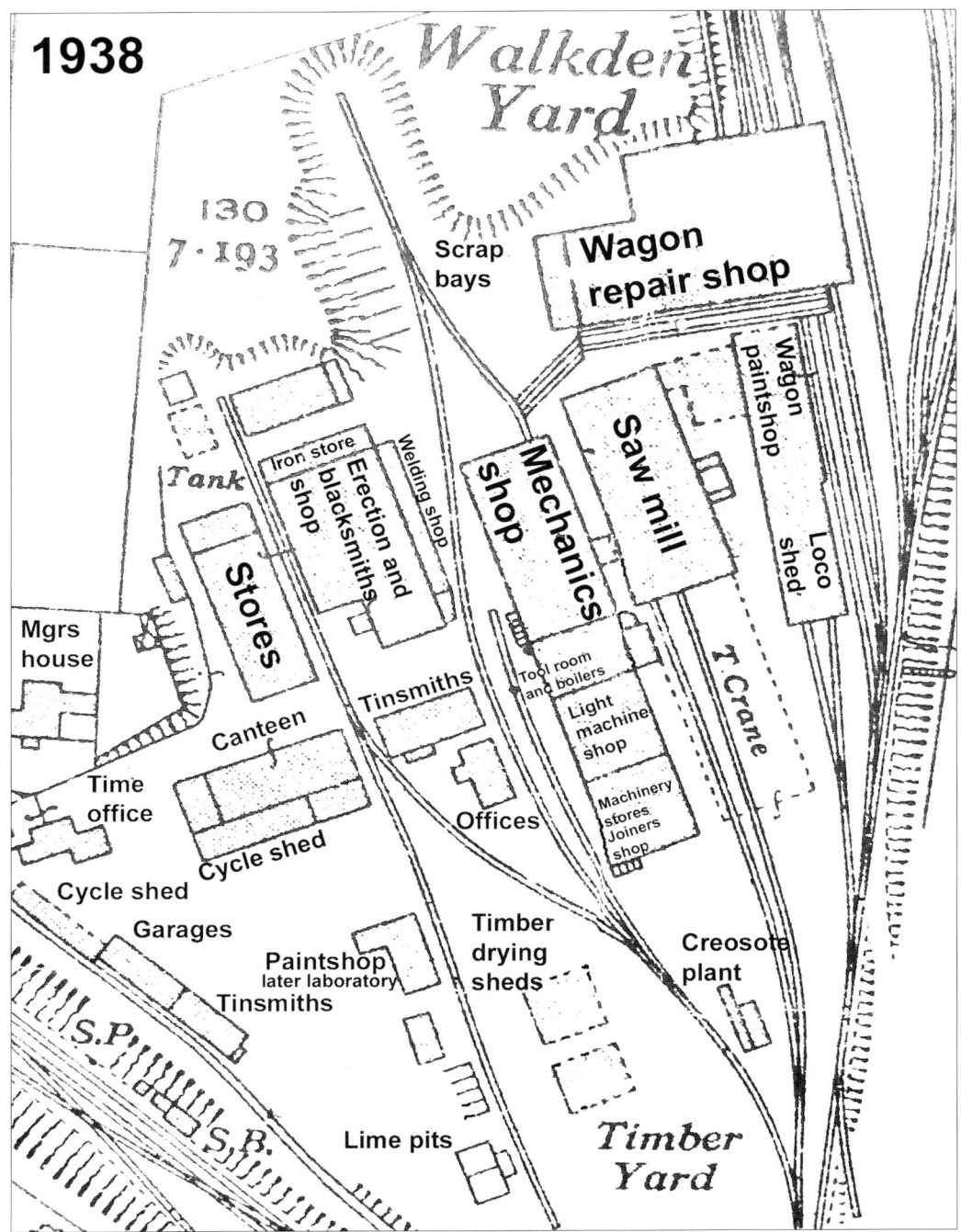

1938

Walkden
Yard

130
7·193

Scrap
bays

Wagon
repair shop

Tank

Iron store

Welding shop

Wagon
paintshop

Loco
shed

Erection and
blacksmiths
shop

Mechanics
shop

Saw mill

Stores

Mgrs
house

Tinsmiths

Tool room
and boilers

Light
machine
shop

T. Crane

Canteen

Offices

Machinery
stores
Joiners
shop

Time
office

Cycle shed

Cycle shed

Garages

Paintshop
later laboratory

Tinsmiths

Timber
drying
sheds

Creosote
plant

S.P.

Lime pits

S.B.

Timber
Yard

Walkden Yard in its late 1930s heyday. The OS plan is dated 1936 and the information overlaid comes from a Manchester Collieries site plan of 1938. The largest building is the wagon repair shop and no wonder – company stock rose from 8,619 in 1934 to 10,966 by 1936, plus internal wagons, creating a large volume of work.

throughout history have had to trust their predecessors' veracity and I also will have to trust George's accounts. Personally I find that George's style does not seem to indicate that he had a grudge to bear so feel they should be included.

Many of the actual working techniques George describes could be found at workshops across Great Britain and more knowledgeable readers involved in present-day restoration work will be aware of this and no doubt be able to spot any implausible accounts!

George Describes the Yard Layout in the 1930s

'So let's start in 1936 by standing in Tynesbank [the road down from the A6 to the Yard] outside the big gates and looking down 'The Yard'. On our left surrounded by a wall and grassy bank stands the house of Mr F. Hilton [Fred Hilton], the manager.

He has, as suits his position a private set of steps from the rear of his house and inside the gates down into the yard, and can from the top of them have a commanding view of people leaving and entering.

He can also see in front of him the Time Office, Garage, Tinsmiths and Fire Station [for many years Fred was to lead the fire team], and by glancing left the Central Stores entrance, Blacksmiths Shop, Paint Shop, and a good part of the Heavy Machines Shop. At least half the yard and any incident occurring in it is under his sharp eyes, merely by stepping out of the house.

To the right of the main gate is the Time Office. Everyone must pass through this both going to and coming from work. Each morning from 6.30 a.m. until 7.00 a.m. Jimmy Denner stands guard at the entrance to the passage through the office, watching keen eyed as each man, and boy, takes his card from the 'In' rack, puts it in one of the time clocks, presses the lever recording time, and puts it into his personally numbered slot in the 'Out' rack.

If you are as much as one second late you can turn back and walk home, you will, under no circumstances be allowed in, losing one days pay, and with the further penalty of having to explain to your foreman the next day your previous days absence!

Walking past the canteen we pass on the left the 'stores'. The home of those superior people, the stores clerks, who required 'Mr' when being addressed, never Jack, Joe and Fred. It is here we cross the first of the railway lines which runs off the 'main line' between the stores and the Blacksmiths' Shop. It terminates in a grassy bank, one hundred yards on near the Electricians' Shop. In the 30s there were two small 0-4-0 locomotives permanently parked at the end of the line, whether they had come to the end of their useful lives, or because they were not man enough to cope with the heavy loads en route to Ashton Field, I never did find out ...'

The Blacksmith's Shop

'Still keeping to the left of the main roadway we have now arrived at 'Hades', the Blacksmiths' Shop. There is a slight rise in the main doors, but already the heavy pounding of the steam hammer just inside and between the long rows of forges on each side of the shop can be felt through the soles of the boots, indeed most of the yard vibrates in tune with it when Bill Yates is forging buffers and wagon hooks under it.

Each of the two dozen or so forges has an electric fan which can be used to control the air and thus the heat to each coke fire. Each forge has in front of it a tank of water, often near boiling point from the repeated quenching of hot metal in it, and alongside it is the standard shaped anvil and accompanying two handed heavy strikers hammer. The floor throughout the whole shop consists of hard earth, rust, scale and coke dust. Whether on piece work, day work, or perhaps even contract work, each blacksmith competes with his neighbour for the traditional two blows on the anvil face before striking his glowing metal into a pick, spade, spike, drag [pit tub arrest bar used below ground] rail for an underground points system, a [pit] cage fitting.

In the case of the other smaller hammers down the centre of the floor, hooks, bolts, tempered chisels, hammer heads, caulking tools and a multitude of parts of pit equipment required for the essential maintenance and continuing production of coal.

The other hammers in the shop consist of a small brother of the Naysmith hammer, also steam operated, a compressed air hammer that clouts the steel in fits and starts, and an electric motor one that emits sparks and flashes as it is worked. They are all skilfully operated by the strikers at the head-nodding of their blacksmiths.

Well down the shop stands the shearing machine. By some mysterious adaptation its blades open and close continuously, so one only has to put metal bars or plates between its masticating jaws for a cut as clean and sharp as that with scissors on paper. It complains with groans and squeaks when it has thick metal thrust into it, but nevertheless does what it is intended of it.'

The Springsmith's Shop

'A wall with doors in it at each end divides the Blacksmiths' Shop from the small Springsmith's Shop, dominated by its brick furnace used for case hardening all manner of jobs. The springsmith and his son work here surrounded by an air of well guarded secrecy. They will not reveal their skills to anyone, and the floor is always a scrap heap of broken loco, wagon and cart springs which will in due course leave their shop re-leaved and well tempered and ready for many more miles of suspension.'

The Heavy Machine Shop

'Crossing the railway lines, much used by the travelling crane, we are now at the wide doors of the Heavy Machine Shop. Through the doors are disgorged small and medium sized, prefabricated, compressed air underground haulage engines. The big ones with up to fifteen foot [diameter] drums are, after construction and testing and having all their parts numbered, painted, dismantled and despatched by lorry to the requisitioning colliery. Out come overhauled disc and jib compressed air coal cutters, vertical boilers for cranes, and those that operate isolated installations; shaker plates [large sheets of metal with graded hole sizes used in the surface coal screens]; large and small shafting; lathe work and plain machined parts.

Standing in the open doorway and looking to the far left, you can see a loco in the process of being overhauled. It is minus wheels, cab and tanks, and the boiler stripped of its asbestos insulation blankets stands high and dry on baulks of timber over the pit and to the rear of the now empty frame. Here it can be de-scaled, inspected, have its firebox re-stayed, its plates re-riveted and caulked, copper fire tubes withdrawn and renewed, and finally water pressure tested.

Only when it has passed the test under the eagle eyes of Charlie Spencer, the boilersmith and the British Engine and Boiler Insurance Co's boiler inspector, is approval given for it to be lifted by the hand operated overhead 5 ton crane and gently lowered into the waiting frame.

If the firebox has been re-stayed with new copper stays, these will have been turned on one of the new turret lathes in the shop, then screwed into the firebox sides until they protrude an equal distance inside and out, prior to having their heads riveted over. This is done first on the outside using a compressed air machine gun chattering [hammer action] riveting hammer. Sammy Pybus on the inside is supporting the blows with his heavy swinging 'dolly'. Harold Whittaker waits to join Charlie when he has finished the initial phase of riveting, then the two of them continue the process each using the hand held special long headed curved shafted riveting hammers, following one another round the head with their blows until they are satisfied it will be water-tight. Each of the several hundred stays are dealt with in a similar manner both inside and outside the firebox.

Alongside the loco is a long bench at which the fitters work. It is here that they inspect and measure the loco's motion work for wear and tear, removing and replacing any worn parts, they scrape new 'Babbit' lined axle bearings until they match perfectly the axles they have to fit. It is here that they prepare their boring bars for truing up the insides of the cylinders, in situ; and it is here that they file and saw, emery and polish, and re-thread bolts, nuts and screws and nail one anothers snap [food] boxes to the bench top.

The Robinson brothers are long experienced fitters and rarely if ever does a bearing run hot after their patient and industrious work has re-assembled the loco. Several fitters will be at work on disc and jib coal cutters, each with its quota of vicious pointed and hardened picks.

To the left of them construction is well ahead on a new haulage engine, probably for Mosley Common [colliery], its drum several feet wide and as high as the boiler firebox. It will be painted a dull grey before it finally leaves the shop.

To the right and left of the wide doors are the two railway wheel lathes, and when being used by Frank White spiral slowly rising, snaking turnings from the revolving wheels, which although innocent looking can inflict severe lacerations on the unwary person who stands too near them.

Further away to the left Stanley Little rides serenely on his long planer platform, machining a long casting. The platform glides back and forth some twenty feet on its bed, between the pillars which carry the cross member and its fixed tool. To the right of the doors in the centre of the shop Bob Galloway operates his extra long lathe, its bed being some fifteen feet in length.

Brilliant blue white flashes light up the white washed wall at the far left of the shop. It is here behind canvas screens that most of the electric arc welding, oxy acetylene cutting and brazing is done. The arc welders draw electric power from a generator next to the boiler house, and set the amount required for the job in hand on ancient upright quasi-arc control sets.

The work range is enormous. Boiler work, tanks, points systems, prefabricated haulage engine sides, conveyor pans and conveyor belt rollers, flanged piping, safety guards of all sizes and shapes, parts of and for three decker [pit] cages, coal cutter pick tipping [for hardening], building up worn mechanical parts, filling up holes for re-drilling, repairing fractured castings, and steel girder construction.

Harry Joyce, the general foreman, has under his command all the work and personnel in the Heavy and Light Machine Shops, including the welders and cutters, platers and riveters, boilersmiths, store and saw doctors. Moving on we have now come into the upper part of the boiler house. Its two Lancashire boilers can be identified by the half circular insulating brickwork which covers their upper halves and by the heat, dust and fumes which arise from the stoke hole. They are maintained and repaired by us, a job none of us like doing. It is the practice to de-scale the fire tubes in the boilers. There are two in each boiler and they extend from the front plate to the back of the boilers. They are about three feet in diameter and about twelve inches apart, except near the front plate.

Manchester Collieries Ltd.

Producers and Distributors of

Quality Coals

Head Sales Office:

Arkwright House,
Parsonage, Manchester, 3.

Manchester Collieries tirelessly promoted the high quality of their coals, analysing in detail their general, steam raising and coking qualities and training sales staff in advising on the variety of uses certain seams were suitable for. Here, 'Emcee Bill' (MC = Manchester Collieries) smiles as he lugs a hundredweight sack of coal to the customer.

MANCHESTER COLLIERIES
LTD.

BUILDING SITES.

We have several plots of freehold land which we are prepared to offer at reasonable terms for sale or on lease

Bolton Road, Pendlebury.

Manchester Road, Kearsley.

Manchester Road and Holden Road, Leigh.

Liverpool Road, near Aerodrome, Barton.

Rake Lane, Clifton.

Salford Road, Middle Hulton.

Tynesbank Road, Little Hulton.

Ringley Road, Outwood.

Land off Church Street, Little Lever.

Manchester Road, Astley.

———

Application should be made to :-
Manchester Collieries Ltd.,
Estate Department, Walkden.

Amalgamating the various coal companies into Manchester Collieries in March 1929 brought together large tracts of land the companies had purchased, either to underwork or develop at some time on the future. The siting of the plots available reflects the constituent Manchester Collieries companies' locations.

The way into the boilers is through the manholes near the upper end of them and once in to get under the fire tubes one has to crawl on them until the front plates are reached, there the tubes narrow sufficiently to allow the average sized person to drop between them. There is just enough head room under the tubes to crawl around with the chipping hammer to remove the scale.

A few paces from the boiler house and we are in the compressor and generator house. The latter is about man enough to cope with the amount of DC current three busy arc welders need, and should they all three 'strike' at the same moment, it will whine and screech as its power is demanded.'

The Joiner's Shop

'Into the joiners shop now, the last of the workshops in this long building. The foreman is Mr Potter, a short stumpy man with a sharp walk. Although there are several woodworking machines on the floor, most of the work is handwork, and there are many benches. A section of the shop is devoted to the wheelwrights. The hub of a wheel is turned in the lathe and then begins the skilled work of shaping the spokes and morticing each one into the hub. With their specialised tools the wheelwrights shape all the segments for the outer circumference fitting each piece in carefully until it is complete. It is then wheeled up to P. Hardman for him to shrink the steel tyre on it.

All the timber used in the Joiners Shop comes from the Sawmills adjacent to the Heavy Machine Shop. It is a huge open ended building, three sides and a roof. The timber logging is brought in by rail and then unloaded by the overhead travelling crane. It is a noisy dusty shop, superintended by Mr Mullineux and a steady stream of planking leaves the floor for use in the railway wagon repair shop next door.'

Wagon Repairs

'Walkden Yard had a very large railway wagon repair shop where damaged wagons could be dismantled and overhauled. The railway wagon repair shop seemed isolated and completely separated from the rest of the yard, as indeed it is. To it come 10 ton railway wagons from all four corners of the country. Some must have been shunted into sidings and forgotten for years. One can only imagine that these wagons had been shunted years previously by one of the main line companies into half forgotten rural sidings and left until some enthusiastic railwayman had decided it was about time they were returned to their owners.

Timber for the repairs come from the Yard timber mills close by the repair shop ready sawn and prepared for use. On one side of the shop all repairs are

Manchester Collieries Limited.   -   Walkden Yard

RAILWAY WAGON REPAIRS

PIECE WORK RATES—JANUARY, 1939

10 TON WAGONS

ITEM	BASIS RATE £ s. d.	PLUS FLAT RATE % £ s. d.
4 Bearing Springs	0 2 0¼	
3 Bearing Springs	0 1 10½	
2 Bearing Springs (opp. ends)	0 1 8½	
2 Bearing Springs (same end)	0 1 0¼	
1 Bearing Spring	0 0 10¼	
1 Laminated Spring, same end as headstock	0 4 1¾	
1 Laminated Spring with Sole Bar	0 4 1¾	
2 Laminated Spring with Sole Bar	0 9 9	
2 Diagonals complete one end	0 7 0	
2 Longtitudes complete at one end	0 6 6	
1 Diagonal and Longtitude at one end	0 6 6	
2 Inter Bar and Cradle	0 1 2	
1 Inter Bar and Cradle	0 1 0½	
1 Pair Pushrods and Brake Block	0 0 6½	
1 Pair Pushrods and Lever and Tumbler	0 0 11	
1 Buffer Castings, same end as Spring	0 0 9½	
1 Y4 Bolt	0 0 7	
1 Y4 Bolt, same end as Headstock	0 0 3½	
1 Diagonal, same end as Buffer Spring	0 2 3¾	
1 Long Bolt with Headstock	0 0 4	
2 Long Draw Bars replaced with Articulated Draw Bars	0 1 9	

RAILWAY WAGONS :—PIECEWORK RATES FOR REPAIRING STANDARD WAGONS. JANUARY, 1939.

ITEM	BASIS RATE £ s. d.	PLUS FLAT RATE % £ s. d.
1 Sole Bar and Planing	0 18 0	
2 Sole Bars and Planing	1 13 0	
1 Sole Bar and 1 Headstock and Planing	1 6 0	
1 Sole Bar and 2 Headstocks and Planing	1 16 0	
1 Sole Bar and 1 Headstock and Planing and 1 M. Bearer at one end	1 17 6	
1 Sole Bar and Planing and 1 Middle Bearer	1 10 2	
1 Sole Bar and Planing and 2 Middle Bearers	1 17 6	
1 Middle Bearer alone	0 16 0	
2 Middle Bearers alone	1 6 0	
1 Headstock and Planing	0 10 6	
2 Headstock and Planing	1 1 0	
1 Headstock and Planing and 1 M. Bearer at same end	1 3 7	
1 Headstock and Planing and 2 M. Bearers	1 12 0	
1 Middle Longitude when Frame open for Middle Bearer	0 1 9	
1 M. Longitude (Frame to be opened specially)	0 12 0	
1 M. Longitude same side as new Sole Bar	0 3 10	
1 M. Longitude opposite side to new Sole Bar	0 5 6	
1 End Longitude (alone)	0 5 0	
1 Diagonal (alone)	0 5 9	

Tortuous, heated and lengthy negotiations must have led to this Wagon Repair Shop pay agreement of 1939 being finally accepted! The Shop was the largest building on site at Walkden Yard and no wonder – company stock rose from 8,619 wagons in 1934 to 10,966 by 1936 (plus internal wagons), creating a large volume of work. Eight months

RAILWAY WAGONS :—PIECEWORK RATES FOR REPAIRING STANDARD WAGONS, JANUARY, 1939.

ITEM	BASIS RATE £ s. d.	PLUS FLAT RATE % £ s. d.
1 Diagonal at other end of Sole Bar to corner opened out	0 4 2½	
1 Diagonal at opposite side to new Sole Bar at end opened out	0 5 0	
Ditto at 'end not opened out	0 5 6	
Additional timbers at ends :—		
Longitude	0 1 8	
Diagonals	0 1 11	
1 End Pillar	0 1 1½	
1 End Pillar (if down for other repairs) ...	0 0 9¼	
1 Side Rail and Packings (including Cross Rods) ...	0 6 0	
1 Side Rail and Packings (including Cross Rods) with Sole Bar	0 3 6	
1 Bottom Plank, new	0 0 2¼	
Bottom Planks, refixing	0 0 1¼	
Ironwork		
1 Buffer Ram (alone)	0 0 9¼	
1 Buffer Casting (alone)	0 1 1¾	
1 Buffer Casting with Ram or Spring	0 0 4½	
1 Buffer Coil Spring (alone)	0 0 9¼	
1 Buffer Check Rubber and Cotter (alone)	0 0 4	
1 Buffer Trimmer (alone)	0 1 8	
1 Buffer Trimmer with Sole Bar or Diagonal ...	0 0 7	

RAILWAY WAGONS :—PIECEWORK RATES FOR REPAIRING STANDARD WAGONS. JANUARY, 1939.

ITEM	BASIS RATE £ s. d.	PLUS FLAT RATE % £ s. d.
1 Diagonal Knee (alone)	0 2 4	
1 Diagonal Knee and Trimmer (together)	0 3 2	
1 Diagonal Knee with new Sole Bar or Diagonal at end not free...	0 1 0	
1 Draw Bar Short alone	0 0 7	
1 Draw Bar, Long with Screw end	0 0 8	
1 Draw Bar, Articulated	0 1 0	
1 Draw Bar, Articulated when replacing long bars ...	0 0 10½	
1 Draw Bar, Intermediate (alone)	0 0 7	
1 Draw Bar Face Plate (alone)	0 1 0	
1 pair Draw Bar connecting Links and Pins (alone)...	0 0 6	
1 pair Draw Bar connecting Links with 1 gedges Draw Bar or Intermediate Bar	0 0 3	
1 Draw Bar Cradle (alone) including Carrier ...	0 1 3	
1 Draw Bar Coil Spring	0 0 7	
1 Headstock and Longitude Knee (alone)	0 1 5	
1 Draw Bar Cradle Carrier	0 0 3	
1 End Door Fastener Bar	0 0 10	
1 End Door Fastener Bar when free	0 0 2	
1 End Door Washer Plate (if new sheeting 3¼) ...	0 0 7	
1 Top through Plank any size (complete job) ...	0 2 8	
1 Top end Plank any size (complete job) ...	0 1 3	

RAILWAY WAGONS :—PIECEWORK RATES FOR REPAIRING STANDARD WAGONS. JANUARY, 1939.

ITEM	BASIS RATE £ s. d.	PLUS FLAT RATE % £ s. d.
1 End Door Knee (if new Headstock 11¾d.)	0 1 3	
1 Axle Guard and Bridle (alone)	0 1 8¼	
Other Ironwork and Sheeting to ordinary list of prices.		
Lifter.		
1 Axle Box bottom half (alone)	0 0 4	
1 Axle Box bottom half when lifted	0 0 0½	
1 Axle Box T. Bolt changing per bolt	0 0 1	
If Roller Bar to remove	0 0 4½ extra	

F. E. HILTON.

after this agreement was drawn up, the Second World War began and wagon repair supplies soon dried up, creating a national wagon shortage crisis, only to be dealt with gradually after the war had ended.

carried out, then by means of an electrical moving platform complete with short rails they are moved across to any vacant spot in the Paint Spraying Shop and after being sprayed a uniform colour [flame red] the familiar 'Manchester Collieries' is stencilled [in white] on each side in bold letters. Sometimes very old 'BT' and BC', Bridgewater Trustees or Bridgewater Collieries wagons would be shunted into the Wagon Repair Shop.

The shop had its own blacksmiths and forges. When repairs had been completed on a wagon it was spray painted and the company's name and details stencilled on it.

All the men in the wagon shop seem a race apart from the rest of the yard. Seldom do they mingle. They keep themselves very much to themselves. Even their accents are different, for they come each morning and return each evening from Walkden High Level Station, having been brought by a local stopping train from the Wigan district. Were they originally Wigan Wagon Co's employees? I often wondered. [In 1933, the Wagon Repairs Co. had bought The Wigan Wagon Company possibly leading to these men losing their jobs].'

The Loco Sheds

'We are now left with only four other buildings to visit. First the Loco Sheds. These are situated behind the wagon paint shop and between the Sawmills and Walkden Offices adjacent to the main Ashton Field – Mosley Common [colliery] line. There are two lines running off the main lines into the shed, terminating over the ash pits. The shed will hold six locos although apart from weekends it is seldom full and even when it is, the rear ends of two of the locos will be well outside the shed.'

Memorable Locos

'I have always remembered the loco names, perhaps because I worked on them and they were our constant companions in the Heavy Machine Shop. They were [virtually] all 0-6-0 locos, others could always be identified by their throaty exhausts, outside cylinders, outside motion work, and side water tanks with cut off front plates. They represented all that one expected a locomotive to look like, especially when they had been newly painted and their brass work well polished.

As for the others; 'Brackley', 'Gower', 'Katherine', 'Ellesmere', 'Cerberus' and one other similar to 'Cerberus' and whose name escapes me at the moment, they were all ugly in design but powerful in action. They were [mostly] six wheelers with inside cylinders, no show of strength unless they were working.

The boilers were capped with large saddle tanks and had squashed up rear coal bunkers. 'Cerberus' [formerly *Atlas* until 1929, 0-6-0STIC Sharp Stewart 2909/1880] and its pal were in a class on their own. They were 'ancients' and that's putting it mildly. Built in the 1860s or thereabouts [1880s], they had tall chimneys, saddle tanks which were only sufficiently wide enough to allow the slimmest and smallest of workmen to enter to de-scale. They had an excuse of a coal bunker on each side of the firebox outside. Their cabs were nothing more than steel plating rising from the front of the 'cab' with two glass plated and brass rimmed eye pieces, suitably obstructed by the saddle tank.

The plating curved over the crews heads and also curved inwards from the sides to a slim piece of metal just enough to protect the driver from a downpour of rain, and I mean down and not driving rain. This curved again to form the backs of the locos.

The [later Austerities and NSR/LMS] locos were capable of hauling 12–14 10 ton wagons of coal from Mosley Common to Ashton Field. Starting from the pit sidings at Mosley Common they were able to gain some speed as they passed the [Ellenbrook] brickworks, then the incline, and the load began to have its effect, and their swaying from side to side, their urgent and quick pull as each cylinder exhausted up the funnel, and a sweating, doubled up fireman spading yet more coal into the firebox were sure signs that by the time the train crossed Hilton Lane [Walkden] engine and crew would be near exhaustion.

At the engine sheds they would come to a staggering halt. The loco would send drifts of lazy, steamy black smoke into the atmosphere, the fireman would renew his efforts to keep some sort of fire going and after a few minutes respite the journey would be started once more, a non-stop run to Ashton Field with the hope and prayer that they would make it without losing too much fire through the fire tubes and up the funnel.

Sometimes the train would be loco – wagons – loco – wagons [actually loco – wagons – loco] all linked together. Then the rhythm and beat of the locomotives' exhausts would change to double quick time, as each engine churned forth its smoke and steam alternatively. The return run was of course quite the reverse, free running all the way downhill often with thirty or more clanging, rattling, tripping braked empty wagons.'

The Paint Shop

'A few short paces from the office is the smallest workshop in the yard – The Paint Shop, with its foreman Mr F. Tyldesley. From its doors flows the multi-coloured, spotless, and freshly painted rubber tyred and wooden wheeled horse drawn carts; the lorries, the barrows, the tarred railings, pitched forks and spades. Every single job that needs a lick of paint has been cleaned and hand painted here. For those jobs too heavy, too big, or too bulky to move into the shop, the painters in their splashed overalls descend on the workshops to carry out their tasks. Paint tins and cotton waste abound on the floor; the smell of musky turpentine pervades the air and brushes by the hundred stand clean and upright on the benches. This is not a place for the oil and grease of the machine shops, or the scale and rust of the Blacksmiths.'

Pay Day

'Friday is pay day and an early finish at 4 p.m. Night workers collect their wages from the time office at 3 p.m. The rest of the employees must queue up at the passage windows, quote their names and number and in return receive a sealed neat brown envelope and payslip. Most men, once paid, stand around the main gates checking their money, paying their betting debts to Charlie Buchanan, or requesting change from those lucky enough to have some. I suspect many

Walkden Yard "A" Team.
Standing (left to right): H. Parkinson, N. Fairclough, J. Tyldesley, J. Hardy, J. May.
Seated   ,,   ,,   J. Denner, L. Bailey (Captain), E. Hilton.

Engineering staff at Walkden Yard are once more victorious in the Manchester Collieries Engineering Department Bowling competition of 1935, seen here in *Carbon* magazine. Established in 1932, the competition was won by Walkden in the first year, then again in 1933, Brackley Colliery winning in 1934. It was perhaps an unfair competition, the Walkden men having their own bowling green at the Bridgewater offices to practice on!

wives never knew the contents of their husbands pay packets. As time drifts by they disperse up and down Tynesbank, to the railway station, the trams and the infrequent latticed-seat buses, collect their bicycles from under the Tinsmiths Shop where the bench held stoves, always burning, heat up the flow of soldering irons to them, and show a warm glow through the windows during the night.'

Overhauling the Locomotives

'In the 1930s the Heavy Machine Shop at Walkden Yard always had one locomotive in it under repairs, and as one was finished and returned to its work another one would be brought into the Machine Shop for overhaul. Before coming into the shop a loco would have had its boiler and tank, or tanks, drained, and most of the coal removed from the bunker (not always!). It would then have been towed up the line between the Machine Shops and the Blacksmiths' Shop by the steam crane at about one mile per hour.

On reaching the extreme end of the line in the scrap heap near the A6 road the points would be changed and slowly, almost at a snail pace, the loco would be pushed by the crane through the big sliding door into the Shop.

The line extended almost the full length of the Heavy Machine Shop and had a deep pit between the rails. Sometimes a saddle tank loco would have to have the tank removed if the loco was having a major overhaul. Then the boiler casing would be removed to reveal the blanket cladding used to conserve the boilers heat. There were several of these blankets, for all the world like huge flock mattresses, almost as thick and just as awkward to handle. It was always a relief when the dusty things had been stored away.

The boiler having now been released from the frame was ready to be removed to the back of it. The boiler was lifted out using the overhead travelling crane. This crane ran on rails almost the full length of the building. The winch on the crane was worked by four men, two to each windlass. It was back breaking tiring work.

Once the boiler was high enough then the men, using another windlass moved the crane and its load along the Shop to its position over the baulks of timber which had been laid over the pit to support it. Once the boiler had settled on to its supports and had been released from the crane this was then travelled back over the frame in case that needed lifting to release the wheels. The extent of the repairs needed to the boiler would be assessed and he had to estimate how much it would cost to do the work, the materials which would be required, and the time it would take. We were paid 'by contract' by the boilersmith! If due to any unforeseen circumstances after the contract had been made, and more repairs were required then it was 'work without pay' or, if you were lucky you went on 'day wage'!

A boiler overhaul and re-fit would possibly have included replacing all the copper stays in the fire box, changing all the copper tubes in the barrel of the boiler between the smoke box plate and the fire box plate for new ones. Building up the wasted parts of the smoke box plate at the front of the engine caused by accumulated ash, smoke, steam and fumes this was done with electric arc welding across the plate between the firebox inner and outer plates and filling the fuse plugs with new lead.

These two last jobs always had to be done. When the riveted steel plates of the barrel overlapped leaking seams indicated by rust on them would have to be caulked up using special caulking chisels and a heavy hammer, with a silent prayer that we had been successful when it came to testing time. We had on occasions to put new foundation rings between the plates at the bottom of the fire boxes. New brick arches, new firebars, all had to be put into the fireboxes and often we had to make new ashpans.

The copper stays were screwed into the firebox joining up the inner and outer plates. The heads of the stays were riveted over both inside and outside the firebox. To remove the old stays the heads had to be chiselled off using a steel chisel held in a pair of blacksmiths tongs and a heavy striking hammer. It was a two man job. The parts of the stays left between the plates were then drilled out using a heavy two man electric drill suspended from chain blocks. Meanwhile new stays had been turned and threaded by a turret lathe operator. These were screwed into the plates with an equal amount of metal protruding both inside and outside the box for riveting over.

The riveting gang consisted of two men and a lad (you guessed right – that was me!). My mate operated the compressed air riveting gun which had at its business end an inner dome shaped head for forming the rivet head. It was my job to turn this head to shape the rivet using a short bar inserted into a hole on the side of the riveting gun while it was hammering away. All the while this was going on outside the fire box our 'brotherly mate' inside the box had the unenviable task of holding a massive steel bar, called a 'dolly' suspended from the roof of the box by a chain, against the stay we were riveting down. When we had completed all the riveting outside the firebox it was our turn to rivet down the heads inside the box. This was an ear splitting session often lasting all day, as you can well imagine.

A stay in a difficult position had to be riveted over with hand hammers. These were special hammers with a long head on one side and a short one on the other, and had a curved hickory shaft to prevent us catching our hands on the boiler plates. It needed some experience to use these hammers.

To remove the copper tubes from the boiler the ends of each tube had to be chiselled away from the smoke box and fore box plates. When the tubes were loose they could be drawn out of the boiler at the smoke box end. Most times this required the Shop sliding door being kept open for some considerable time much to the disgust of the fitters and machine operators, especially on a cold, wet day in winter. New tubes would have already been obtained the required length for the boiler but before they could be put into it the ends of each tube had to be softened by heating them in our forge to a dull red then plunging them into water. The tubes projected a fraction of an inch outside the plates and I have often been asked how they were made steam and water tight.

The holes in the plates were slightly larger than the diameter of the tubes. We had a special tool. It was some twelve inches long and consisted of three tapered rollers. These three rollers were held together in the form of a triangle by a plate at each end. Through the centre and between the rollers was another one which could turn the outside ones as it had a squared end to which could be fitted a large spanner. The rollers were placed inside a tube, and tapped in until they were tight, then they were turned with a spanner to expand the tube onto the plate. At frequent the rollers were tapped farther into the tube and rolling continued until the rollers were completely in the tube. This made the tube ends both water and steam tight when under pressure. When all this work had been completed to our boilersmiths satisfaction it was time for the water pressure test. This always had to be done in the presence of a certified Board of Trade boiler inspector employed by the British Boiler Engine and Electrical Insurance Co. of Fennel Street, Manchester, the collieries' boiler insurers.

The boiler was filled with water and a special fuse plug without its lead core was connected to a hose pipe and pump and screwed into the boiler for building the pressure up to around 150–180 lbs per sq inch and kept at these figures while the inspector and the boilersmith checked every part of the boiler and all the work we had done on it for leaks.

On one occasion when testing a loco boiler the fuse plug and hose pipe worked loose and were forced out of the outside plate of the firebox by the pressure we were putting into the boiler. A jet of water some two inches in diameter shot across the Heavy machine Shop like a long steel bar at a great speed and at waist height. Very fortunately for all concerned no one was hit by it otherwise there would have been a fatal accident. When the inspector was fully satisfied with the test the pressure was slowly released and the boiler was ready for lifting and putting back in the frame.

Locos sometimes had a firebrick arch collapse, or the firebars would burn together while it was out working on the mineral line. When this happened my mate and I would be instructed to report and 'clock on' late the same evening. We would find the 'disabled' loco parked over the pits in the loco sheds. The fire would have been dropped out and the steam run off. New firebricks, cement, and new firebars would be alongside the loco ready for us to put into the fire box when we had removed the remains of the old and damaged ones. It was a thankless, fatiguing and miserable job, despite us receiving overtime pay for doing it. Many times the interior of the firebox, the broken bricks and firebars were still hot. We usually tossed a coin to decide who would 'have the honour' of crawling through the fire hole into the box first. I am sure my mate had a double headed penny, for I always seemed to lose the toss, and have to be the first into the inferno. With heavy duty leather gloves on my hands, and with little else, even in winter, although still respectable, the remains of the brick arch would be man-handled through the firehole onto the footplate, and from there into a barrow for later disposal.

The bricks were made out of fireclay and were slightly curved lengthways, about four to five inches square and about two feet long, a considerable weight to handle within the confines of a firebox. Each brick could be interlocked into the previous one forming an arch below the bottom row of copper tubes. By the time the last piece of broken brick had been removed my mate and I had probably changed places several times giving each of us time to cool off. The new bricks had to be lifted almost head height from the side of the loco and onto the footplate and stacked alongside the fire hole, then man-handled one at a time into the firebox and cemented into the interlocking positions very carefully.

The fire bars which were badly burnt, broken or fused together frequently had to be prised loose with a heavy hammer and a crowbar, and where they could not be dropped into the ash pan, lifted and pushed through the fire hole. At some point in their removal you had to stand in the ash pan for the few remaining bars. Those dropped into the ash pan had to be removed by jumping down into the pit, dragging them clear of the ash pan, and lifting them up clear of the side of the loco.

By the early hours of the morning if all the work had gone smoothly and the job had been completed we would compensate ourselves with a mug of strong tea (we never had time for a snack) find some water for a wash, put on our 'going home' clothes, clock off at the time office in Tynesbank near the railway bridge, struggle down the banking onto the main line, and quietly make our way between the two sets of tracks homewards, too exhausted for any conversation.

Talk could wait until later in the day when we again clocked on at twelve noon, by which time we would probably hear and see our nights work thundering its way past the Yard and Sheds to Ashton Field. Our entry into the Heavy Machine Shop would be met with ribald comments, and sly assertions about oversleeping and being too late to clock on earlier in the day, when the shift had started at 700 a.m. Still more or less half asleep, and not having fully recovered from our nights work, we would report to the Shop Office, hand in our time sheets and ignore scurrilous remarks made by all and sundry.

It was natural when working on a boiler, when time allowed, to watch the work being done by the fitters on the frame, the wheels and the motion work.'

Retyring

'If the wheels required new tyres they had to be removed from under the frame. The travelling crane was used to lift the frame up, and when it was sufficiently high enough the axles and wheels still on the rails were pushed from under the frame through the sliding door, along the track and down to the front of the Heavy Machine Shop. Here we would be involved in cutting off the worn tyres using oxyacetylene equipment.

New tyres would have already been obtained, and after meticulous measurements had been taken of the diameter of the wheels, the tyres one at a time were secured in the wheel turning lathe, and the inside of each one turned out until it was a few thousandths of an inch less in diameter than the wheel it was going to be fitted to.

The tyre was then taken outside to a blacksmith and his striker. Outside the Blacksmiths Shop close by the railway line the tyre was supported on fire bricks and a wood fire lit all the way round and underneath it. The blacksmith would frequently check the tyre's expansion using a mark he had put on it, and a mark on a small wheel held in a fork which he would run round the inside of the tyre.

From long experience he would know when the tyre had expanded enough for the wheel to go into it. Meanwhile the steam crane would have already have upended the wheels and axle with the wheel to be re-tyred at the lowest point. At a signal from the blacksmith the crane would slowly lower the wheel into the tyre, ably assisted by the blacksmith and his striker with their heavy two handed hammers, until the wheel rested against the shoulder inside the tyre. The tyre and the wheel would then be hosed down with sufficient water for shrinking to occur as quickly as possible. When both wheels on an axle had been re-tyred the whole assembly was taken into the Heavy Machine Shop, and there secured in the wheel turning lathe and the tyres turned up true.

A locomotive with cylinders inside the frame would sometimes need them re-boring. This was a major job for the fitters who used a boring bar with a tool fitted on it and a heavy electric drill suitably geared down to a slow speed, and securely fixed to the frame. The pistons then required new rings fitting to them. The wheel bearings were made of hollowed brass filled with white metal, not unlike lead or solder. The old worn white metal was melted out of the two halves of a bearing, and with additional metal added they were refilled, then the complete bearing was turned up on a lathe until it would fit the axle.

This was done on all the wheel bearings of a loco. When each bearing had been thus treated, they were returned to the fitters for their final finish and individual treatment. The fitters used 'Engineers Blue' paste on each axle, and then by rubbing the bearing on the axle the 'Blue' would highlight the parts which needed further attention, these parts were dealt with by scraping the white metal until it perfectly matched the axle which was to run in it.

There were many other repairs and replacements which had to be done by the fitters to the motion work, pistons, steam valves, cab controls and brakes. Sometimes a leaf spring, or perhaps all the wheel springs had to be removed from the frame and sent to the Springsmiths who had their own forge and furnace in the Blacksmiths Shop. Broken spring leaves would be replaced, loose rivets tightened up in them, tempering done, even new springs constructed. Loco springs took a great deal of punishment due to the constant strain they were subjected to.

When a locomotive's repairs had all been completed, and it had been rebuilt, it was the job of the painters from the Yard Paint Shop to come and re-paint, and 'line-out' the engine until it looked very smart and dare one say – almost new. Finally with a rub over all the exterior brass work and name plates the locomotive would be moved out of the Heavy Machine Shop, fired up and returned to the Sheds and its crew.'

In Conclusion

'I have no doubt in my mind that the efficient and long periods of work these engines did over the years without being cosseted or having excessive attention was in no small measure due to the long experienced and skilled work of the fitters, blacksmiths and others at Walkden Yard, and also the crews at the Sheds, even more so when one realises that on which ever section of the mineral line the locomotives were working they constantly faced much uphill work with considerable loads.

These industrial locomotives, which hadn't the attractive, sleek design or eye appeal of the main line railways stock were as familiar to the inhabitants of Walkden, Little Hulton and Worsley as the local trams and buses, even though they produced much noise, steam and smoke at all times of the day and night.

And what of Walkden Yard itself? Over the years as the N. C. B. closed the repair facilities at Haydock [formerly Richard Evans & Co Haydock Collieries], and Kirkless, Wigan [formerly Wigan Coal & Iron Co Ltd later Wigan Coal Corporation headquarters] and [with] further reorganisation of areas Walkden Yard became responsible for the supply of locos to the whole of the Lancashire Coalfield, Cumberland and North Wales, and for their maintenance. Even two locos from South Wales came to Walkden for overhaul.'

Nostalgia?

'You may well ask having now reached this point in my account 'Did you enjoy working at the Yard?' That is a difficult question to answer. One looks back on the days there with some nostalgia. In the days of much unemployment, and paltry dole it was enough to have a job and security, and there was security at Walkden Yard. I never remember anyone being sacked.

The wages were reasonably good (although my first weeks wage was the princely sum of 5s 2½d a week – 27p) and later as I became more skilled I could earn between £2 and £3 per week especially on piece work. We worked a five day week, later to be increased to five and half days.

There was a great variety of work, particularly on the welding and cutting side. We often went out to do repairs at Newtown Colliery, Sandhole Colliery, Mosley Common Colliery, Brackley Colliery, Ellesmere Colliery, Bradford Colliery (near Manchester), Ashton Field (tub repairing), Astley Green Colliery, Chanters Colliery and many others in the combine.

Most of the men I worked with were really rough 'diamonds'. They were friendly, cooperative, loyal and hard working. They knew their jobs intimately and the skills required. Their language and conversation could be, and often was coarse, sometimes bordering on the obscene. They were always ready for the joke and trick, and would enjoy immensely anyone's discomfort at the same time if only to break the rhythm of the work in hand. I doubt whether they aspired to any greater heights than they had already achieved, they always seemed content with their lot, and apart from an odd one or two, must have worked at the yard from leaving school, and would do so until they reached retirement age.

Occasionally I would meet retired workmen who had come to the time office for their weekly pension. It was by todays standards a mere pittance of 5s – 25p per week. Even so that small amount was half as much again as the state pension

A rare and unissued Manchester Collieries Ltd alloy pay tally dating from around the mid-1930s. A man's three digit works number would normally be stamped in the centre. After stating his pay number and collecting his tally from the time office, he could collect his wage at the cashier's window.

and must have made them feel well off compared to those who only received the state pension of 10s [50p] per week. I have no recollection that we ever voluntarily contributed to the collieries pension scheme.

I don't think any employees at the yard felt they were the inheritors of the traditions of the early pioneers of the East Lancs coalfields. The yard was just a secure place of employment, and as long as you kept your place, behaved yourself, and carried out your duties, you could consider you had full employment for all your working life. The demand for coal was increasing throughout the late 1930s, of that I have no doubts, otherwise we should not have been kept in full employment. It was our job to support the expanding mines, as and when necessary. So I think my answer to the question really ought to be – to a certain extent, I learned the hard way what a working mans life was like in the 1930s but that knowledge was of immense value later in life. I was contented – there was little else to do'.

An idea of the locomotive throughput at Walkden can be seen in the figures for year ending March 1935, when fifteen steam locomotives had been overhauled. It should be noted that Manchester Collieries also had extensive workshops at Gin Pit, Tyldesley, where colliery mechanical equipment and locomotives were repaired. Gin Pit dealt with locomotives from the former Atherton Collieries of Fletcher Burrows & Co. along with Astley & Tyldesley area locomotives.

Purchasing Power

Manchester Collieries' director Sir Robert Burrows (born 17 March 1884) was also a director and later the last chairman of the LMSR. It was then decided, not surprisingly, by the board of Manchester Collieries to buy five 0-6-2 tank locomotives from the LMSR. Four of these New L Class locomotives had been constructed at the former Stoke works of the North Staffordshire Railway between 1913 and 1923. The first (NSR No. 221, LMSR No. 2264) reached Walkden on 12 June 1936, to be named *Kenneth* (after Miles Kenneth Burrows formerly of Fletcher Burrows & Co. Ltd). By 8 October 1937 the remaining four locomotives, to be renamed *Sir Robert*, *King George VI*, *Queen Elizabeth* and *Princess*, had arrived.

Interestingly, at this time locomotives based at Walkden Yard were coaled at Mosley Common Colliery directly from pit tubs on an overhead gantry. What rank of coal was used is not recorded; possibly the ultimate steam coals, Trencherbone or Rams, were utilised – certainly other seams such as the Brassey or Plodder would not be suitable, being of middle rank, dirty and high in ash content.

In 1938 Manchester Collieries consisted of the following:

Colliery	Location	Underground	Surface
Ashtons Field	Little Hulton	1	26
Astley Green	Astley	1,402	575
Nook	Tyldesley	932	290
St Georges	Tyldesley	505	149
Shakerley 'Nelson'	Tyldesley	129	100
Gin Pit	Tyldesley	293	227
Bedford	Leigh	771	307
Chanters	Atherton	876	447
Gibfield	Atherton	474	283
Howe Bridge	Atherton	267	132
Ellesmere	Walkden	0	5
Mosley Common	Tyldesley	1992	612
Newtown	Clifton	585	236
Wharton Hall	Tyldesley	2	3
Brackley	Little Hulton	650	206
Sandhole	Walkden	941	373
Wheatsheaf	Pendlebury	644	290
Bradford	Bradford, East Manchester	487	248

Fred Hilton, Walkden Yard Manager 1924–50

I interviewed Bessie Hilton, aged ninety-one, daughter of Fred Hilton, in December 2012, Bessie lived in the yard manager's house from the age of three (1924). Her father Fred was born in Boothstown in 1884. He was to become probably the most influential and skilled colliery engineer in control of the yard during its lifetime. He was in the post at the time of the formation of Manchester Collieries in 1929, and through the difficult 1930s period of rationalisation and standardisation of colliery working methods, equipment and marketing. He saw Manchester Collieries rise to become one of the largest and most profitable colliery combines in Britain, a company needing the highest standard of engineering backup and expertise.

After the coal shortages and manpower problems the coal industry had faced in the Second World War, with Walkden Yard having to 'do its bit' alongside its mainstay work, he witnessed the nationalisation of the industry in 1947. Sadly he died after an arduous site visit to a colliery only months before his retirement on 27 January 1950.

What was Fred's background before coming to Walkden Yard as master of the yard in 1924?

He had been apprenticed to Bridgewater Collieries at the old Worsley Yard [closed by mid-1900]. He studied colliery engineering at technical college, gaining experience by working at Brackley Colliery, then Sandhole Colliery. I was three by the time we moved to Tynesbank House [the manager's house at Walkden Yard]. I was born at Greenheys [close to Brackley Colliery, Little Hulton], the youngest of three daughters.

Father had been based at collieries until the post at Walkden Yard became available; the position at that time was known as master of the yard.

In this photograph Fred appears to be in uniform.

That was the fire brigade leader's uniform. There wasn't a local fire brigade at Walkden apart from the one that had been set up by the Earl of Ellesmere at Bridgewater Offices, then Walkden Yard [Fred took charge of the brigade]. There was always a driver on duty at the yard, even through the night. The driver would get the engine out, the whistle would be blown at the bottom end of the yard and the fire crew would hear this and arrive at the yard or be picked up nearby.

Living in the yard master's house did you get the chance to wander about the yard and chat to the workmen?

Yes, many times, but we never went unless we were with our father. To go into Walkden we used to walk through the yard and up Mayfield Avenue instead of walking up Tynesbank.

Proprietors of
COLLIERIES (CLIFTON & KERSLEY) LTP
OUTWOOD COLLIERIES LTP
AND
PILKINGTON COLLIERY CO LTP

Coal Sidings:
CLIFTON. UNITY BROOK.   L & Y.
"   ROBIN HOOD.   L & Y. & L & N.W.
OUTWOOD. RINGLEY ROAD.   L & Y.
ASTLEY. ASTLEY GREEN.   L & N.W.

TELEGRAMS: "FUEL PHONE. MANCHESTER.
TELEPHONE: 4331 CENTRAL.

MANCHESTER OFFICE. }
28. CROSS STREET. } 782 CITY.

DEPÔTS:
MANCHESTER & DISTRICT.

TRAFFORD WHARF.
OLDFIELD WHARF.
LIVERPOOL STREET.
BESWICK.
MILES PLATTING.
 (CLIFTON ST.)
FALLOWFIELD.
LEVENSHULME.
ALTRINCHAM.
SALE.
ECCLES.
ALDERLEY EDGE.
WILMSLOW.
CHEADLE HULME.
CRUMPSALL.
MIDDLETON.
PRESTWICH.
BOLTON.
OXFORD STREET.
HALLIWELL SIDING.
BLACKBURN.
DAISYFIELD SIDING.
BURY.
BOLTON STREET.
BUCKLEY WELLS.
RADCLIFFE.
JAMES STREET.
LIVERPOOL.
CROWN STREET.
RATHBONE ROAD.

THE CLIFTON & KERSLEY COAL CO LTD
CLIFTON,

Ref: FAW/N.

Near MANCHESTER.

Railway Station:
DIXON FOLD. L & Y. RLY.   9th March. 1929.

Mr. Fred Hilton.
 Walkden Yard.

This is to inform you that you have been appointed the Manager of Walkden Yard and Ashton's Field Repair Shops under Manchester Collieries Limited.

When Mr. Gibson takes over control of the Central Shops of the Company, which will take place shortly, you will come under him.

In addition to the above duties you will temporarily act as Enginewright at Ashton's Field and Ellesmere Collieries.

A Memorandum dealing generally with the responsibilties and duties of Enginewrights, etc., and on the organisation of the Engineering Department, will be sent you shortly.

Your salary will remain as at present.

Manager Fred Hilton is formally notified that he will carry on in his role after the formation of Manchester Collieries, with an additional note that his wage will stay the same! (Bessie Hilton)

Left: Fred Hilton, Walkden Yard manager, is seen here in his fire chief's uniform in the early 1930s. Colliery engineer Fred was already working for Bridgewater Collieries when he was promoted to the post at Walkden Yard in 1924. (Bessie Hilton)

Below: Fred Hilton had replaced the old Bridgewater Collieries Merryweather type fire engine soon after he arrived as Master of the Yard in 1924. Here he stands with his volunteers in the early 1930s, the engine about to join the Walkden Carnival procession. Fred continued his enthusiasm for fire service work during the Second World War, operating the Auxiliary Fire Service at Walkden Yard with volunteers. (Bessie Hilton)

The manager's house seems quite large in your photo of 1940.

Tynesbank House was quite substantial; it was identical to the one on the East Lancashire Road – Woodside, a colliery house [Woodside was the Mosley Common Colliery manager's house, now a pub/restaurant]. You never missed a bus or a train because the office [Bridgewater Offices] clock told you the time.

I think my parents would have loved a boy – I was the third girl, and I spent a lot of time in the yard, I can remember collecting the little curls [swarf or turnings] from the blacksmith's shop. I was allowed to pick them up, provided I picked them off the floor. We would walk through the yard four times a day [when at school]. The men were very friendly and would have a laugh or smile at us, we knew who was who.

In all the time you lived at Walkden Yard did you ever get the chance to visit a coal mine?

Tynesbank House in January 1940. Master of the Yard Fred Hilton moved in here in June 1924. Identical to the Bridgewater Collieries, Mosley Common Colliery manager's house today survives as a pub and restaurant alongside the A580 East Lancs Road. (Bessie Hilton)

Yes, we went down Mosley Common Colliery in the 1940s but didn't go all the way to the coalface.

The canteen at Walkden Yard arrived in 1932; what did the men do before then?

They used to cook their bacon on the forge fires in the blacksmith's and mechanic's shop.

Did men travel from various locations to work at Walkden Yard?

No, they tended to be local men, the majority walking to work. A few came from Farnworth. When there was a lot of unemployment [in the late 1920s to early 1930s] men used to wait at the gates for my father in the hope of asking for work. It was awful and my father hated it so he made another way out of the house into the yard to avoid having to tell them that there wasn't any work.

Did Fred stay at the yard most of time or was he called out to collieries?

He was called out occasionally to advise at collieries and also at canal barge workshops, as well as the Ship Canal, Mode Wheel Locks near Irlam.

What about the wartime period?

In the Second World War father set up the Auxiliary Fire Service brigade, staffed by volunteers and based at Walkden Yard. This later became the NFS, National Fire Service. During the war the Ministry of Supply had them making track pins [for tank tracks].

Do you think your father enjoyed his time at Walkden Yard?

Oh yes, he loved his job, he'd gone out to work before we had our breakfast in the morning, because the yard started at 7. The meetings of the Manchester Collieries' managers and directors were held at the Bridgewater Offices, and occasionally we met people such as Sir Robert Burrows and Miles Kenneth Burrows. Looking back I think my father's greatest skill was as a manager of men, a person who could get a team of men working together to solve a problem.

On the day of his funeral Walkden Yard stopped work, the men lined Tynesbank all the way up. He had always been concerned for the men's welfare and was highly respected, as was his wife.

NATIONAL COAL BOARD
NORTH-WESTERN DIVISION

TELEPHONE:

TELEGRAMS:

OUR REF:

YOUR REF:

Temporary Address:

47 Peter Street,
Manchester, 2.

29th January, 1947.

Dear Sir,

I have pleasure, on behalf of the North
Western Divisional Board, in conveying to you
thanks and appreciation for the able way you have
performed your varied duties during this most
difficult month.

The progress made has been most marked,
and I feel sure, given the continued co-operation
of everyone concerned, the success of the Industry
in this Division is assured.

Yours sincerely,

[signature]

Chairman.

Mr. F.E. Hilton,
Tynesbank House,
Little Hulton.

The newly formed National Coal Board NW Chairman, James Webb, thanks Walkden
Yard manager Fred Hilton for all his hard work in the first month. Loco *James* (7175/1944,
WD 71521) Robert Stephenson & Hawthorns, was in return named after Mr Webb after
its purchase by the NCB in June 1947. (Bessie Hilton)

The War Years and Aftermath

After the First World War many export markets for British coal were lost. The industry slumped into a depression which was to last nearly until the onset of the Second World War. The purchase of new locomotives by mining companies declined, with many locomotive builders going out of business.

The engine shed at Mosley Common Colliery operated until the outbreak of war when the work was transferred to Walkden Yard due to manpower shortage. During this period Walkden Yard went into 'make do and mend' mode, no new locomotives being added to the fleet. Large industrial sites had been highlighted on Luftwaffe bombing maps of Britain, including Kirkless Iron Works at Wigan, Mosley Common Colliery and Astley Green Colliery. Being adjacent to the Bridgewater Canal, Astley Green was a prime target with the reflecting water feature helping to guide navigators and bomb aimers, especially in moonlight. Bombs were dropped in August 1940, luckily missing Astley Green, landing south of the canal. During Christmas 1944 a V1 flying bomb landed on a house at Worsley Dam, near Worsley Delph. Manchester Collieries' chief chemist, Mr Guider, was worst affected by the blast, but company director F. A. Willink (previously the company's engineering executive officer) also suffered some damage.

In wartime basic supplies for sites such as Walkden Yard were suddenly very hard to come by. Work on machining tank turrets and making tank track pins had been farmed out to Walkden, echoing the munitions work carried out there in the First World War. Work was also carried out on metal-framed structures, the men having no idea what their purpose was. The confidentiality was relaxed after the D-Day landings, and the men were told that they had been constructing Mulberry Harbour sections. During this time an armoured vehicle (possibly a prototype) was also constructed at Walkden Yard. It proved to be disastrous, underpowered and too heavy, and was not able to make it to the top of Tynesbank!

During the Second World War the Luftwaffe had a series of roughly 6 inch (1:10,560) to a mile maps marked with target industrial sites such as Wigan Coal & Iron Co.'s Kirkless Iron Works and Mosley Common Colliery, easily located by the East Lancs Road. They would have been well aware of the importance of Walkden Yard and the Bridgewater system. Air raid shelters were constructed at the collieries soon after the outbreak of hostilities. Walkden Yard's shelters were still on site after closure and are seen here in summer 1987. During the war, shelters were also situated behind the Bridgewater Offices. (Alan Davies)

Hunslet's standard works photograph of their classic Austerity locomotive, designed from 1942–3, here seen in their 1946 parts catalogue, copies of which were still on site at Walkden in 1986.

An interesting group in the Manchester Collieries February newsletter. From left to right: W. H. Richards, Mining Agent, Central District (loco *WHR*); Gordon Nicholls, Mining Agent, Western District (loco *Gordon*); E. Humphrey Browne, Production Manager (loco *Humphrey*); and Allen Beaumont, Mining Agent, Eastern District (loco *Allen*).

In 1942 the Ministry of Supply aimed to promote the design of a simple yet tough and powerful shunting locomotive. In 1943 they announced, 'Orders have been placed with locomotive manufacturing firms for a number of 0-6-0 saddle tanks to a simple and robust design based by the Ministry on a standard shunter of a well-known locomotive building form.'

Thus began the career of the famous Hunslet 0-6-0 saddle tank Austerity locomotive. The ministry insisted on a product which could be speedily constructed by fabrication, welding and the use of alternative materials whenever possible. The specification allowed for a hard-working life of only two years.

This was to see service in many parts of Europe, proving so successful that 485 were built in total. The situation regarding the railways during the war became desperate, especially the national shortage of main-line and internal wagons. The government had taken over control of the railways in September 1939, requisitioning all privately-owned main-line wagons to create a national pool, along with railway-owned wagons. The owners were paid hire compensation.

Times were tough. Manchester Collieries was to be one of only a small number of the giant colliery merger companies in Britain to see its equity value decline by 1943–44. On a local level one example of the stressed financial and supplies situation was the use of motor lorries (from 1944) at Astley Green and Mosley Common collieries to remove pit waste.

After D-Day (6 June 1944) many WD 0-6-0ST Austerity locos were sent to the Continent until early 1945. In Belgium they were used mostly at the

Production of the powerful and efficient 0-6-0 WD Austerity design, based on a design of 1923, began in 1943 and continued until 1964. By 1947, 377 had been built for the War Department. As the final batch of War Department locomotives was being delivered, the National Coal Board orders came in. Between 1948 and 1964 seventy-seven new Austerity locomotives were built for the NCB.

This Hunslet advertisement appeared in the 1947 edition of *Guide to the Coalfields*, the first after nationalisation. Their rugged 100hp 0-6-0 flameproof diesel underground locomotive became a familiar sight at most collieries.

HUNSLET

first in the field and still leading

More "Hunslet" designed 0-6-0 Austerity tank engines are now building in our works for the National Coal Board. We also provide a full spare parts service for these locomotives.

The HUNSLET
100 h.p
DIESEL MINE
LOCOMOTIVE

is Britain's first and foremost heavy duty unit and has been in large scale production for a considerable time, giving reliable and economical underground haulage service in collieries all over the country.

The first 100 h.p. flameproof loco in underground service.

THE HUNSLET ENGINE COMPANY LIMITED
HUNSLET ENGINE WORKS   LEEDS 10

two Antwerp dock sheds, Antwerp Dam and Antwerp South. After the end of the war in Europe (the German instrument of surrender was signed on 7 May 1945) engines went into store at Calais and Antwerp for about a year, prior to disposal. Some locos were sold to France, Tunisia and the Dutch State Railways and interestingly a small number were loaned to the Dutch State Mines. The LNER purchased seventy-five, and they became the familiar J94 Class; the War Department retained ninety of the locos for their various depots. Most of the locomotives were to be disposed of in 1946 and 1947, and the majority of these were stored in England and left War Department service in 1946; those in store abroad left service in 1947.

The UK coal-mining industry passed into nationalisation during this period (January 1947) and welcomed with open arms these tough, powerful locomotives. The range of Austerity 0-6-0ST locos present in the industry at this time ranged from those loaned new to various collieries when built, those purchased by coal companies from WD storage in Britain in 1946, locos purchased by the new National Coal Board from store abroad in 1947 and locos operated by the Directorate of Opencast Coal Production of the Ministry of Fuel and Power. The opencast locos were frequently transferred from site to site, at times housed in nearby colliery engine sheds.

In late 1944, surplus-to-requirement WD Austerity locomotives from the Longmoor Military Railway in Hampshire found their way to Walkden Yard, the first being WD 71501 (later *Charles*), the second WD 71500 (later *Allen*, after Allen Beaumont, Eastern District Mining Agent) arriving on 19 April 1945.

After the war had ended Manchester Collieries purchased three more WD engines: WD 71479 (later *Gordon*, after Gordon Nicholls, Western District Mining Agent), WD 71484 (later *Humphrey*, after Humphrey Browne, the former Manchester Collieries production manager) and WD 71480 (later *Fred*, after Fred Hilton, long-serving manager at Walkden Yard, 1924–50).

The Hunslet Austerity Locomotive, a Summary

Original designer	The Hunslet Engine Company, Jack Lane, Leeds
Builders	Hunslet Engine Company (228), Andrew Barclay Sons & Co. (15), W. G. Bagnall (52), Hudswell Clarke (50), Robert Stephenson and Hawthorns (90), Vulcan Foundry (50)
Build date range	1943–64
Total built	485
Configuration	0-6-0ST
Gauge	4 feet 8 1/2 inches (1,435 mm)
Drive wheel diameter	4 feet 3 inches (1.295 m)
Minimum curve	180 feet (54.86 m)
Wheelbase	11 feet 0 inches (3.35 m)
Length	30 feet 4 inches (9.25 m)
Axle load	13 tons 7 cwt (29,900 lb or 13.6 t)
Loco weight	48 tons 5 cwt (108,100 lb or 49.0 t)
Fuel capacity	2 tons 5 cwt (5,000 lb or 2.3 t)
Water capacity	1,200 imp gal (5,500 l, 1,400 US gal)
Boiler	Round top outer firebox, 181 tubes, copper or steel inner firebox
Boiler pressure	170 psi (1.17 MPa)
Fire grate area	16.8 square feet (1.56 m2)
Heating surface, tubes	873 square feet (81.1 m2)
Heating surface, fire box	88 square feet (8.2 m2)
Cylinders	Two inside
Cylinder size	18 in × 26 in (457 mm × 660 mm)
Valve gear	Stephenson
Valve type	Slide valves
Tractive effort	23,870 lbf (106.18 kN)

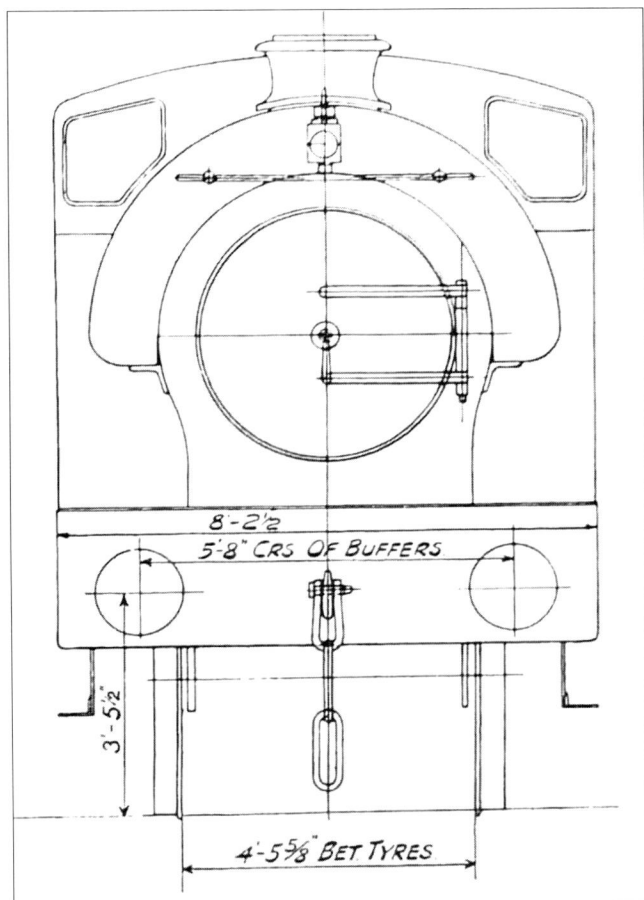

Right: Wartime Hunslet Austerity parts catalogues in various states of disintegration were still to be found at Walkden Yard on closure (they have since been superbly reprinted by Camden). This frontal illustration comes from a 1946 edition.

Below: 1946 Hunslet Austerity parts catalogue.

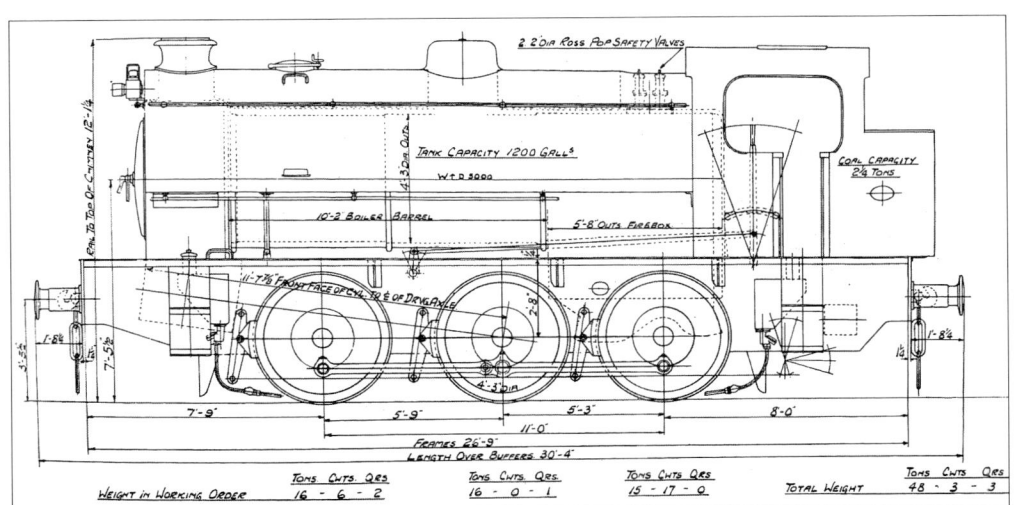

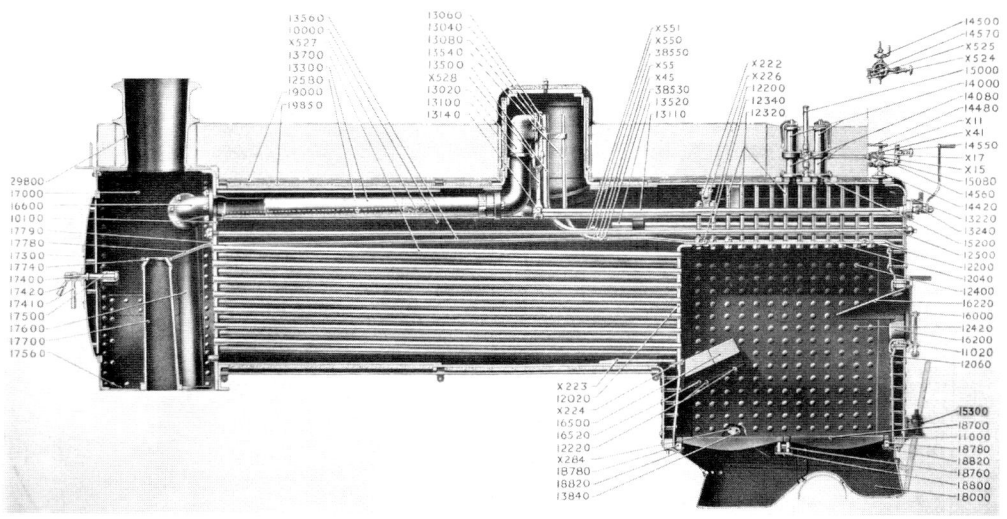

1946 Hunslet Austerity section illustration.

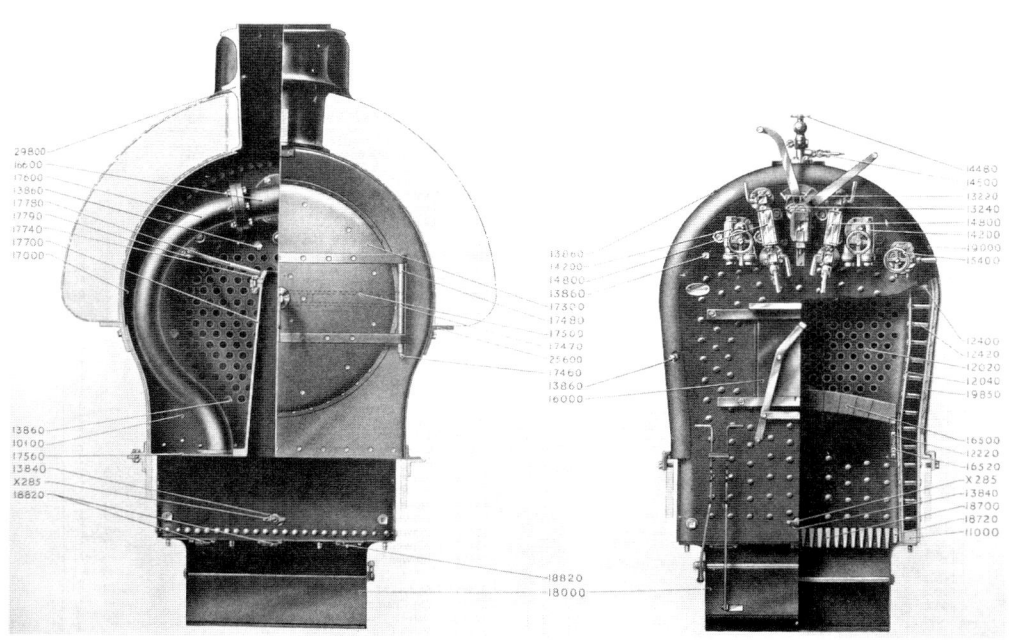

1946 Hunslet Austerity parts catalogue, smoke and fire box sections.

Opposite above: 1946 Hunslet Austerity parts catalogue, slide valve arrangement.

Opposite below: 1946 Hunslet Austerity parts catalogue; cylinders, connecting rod, cranks and wheels.

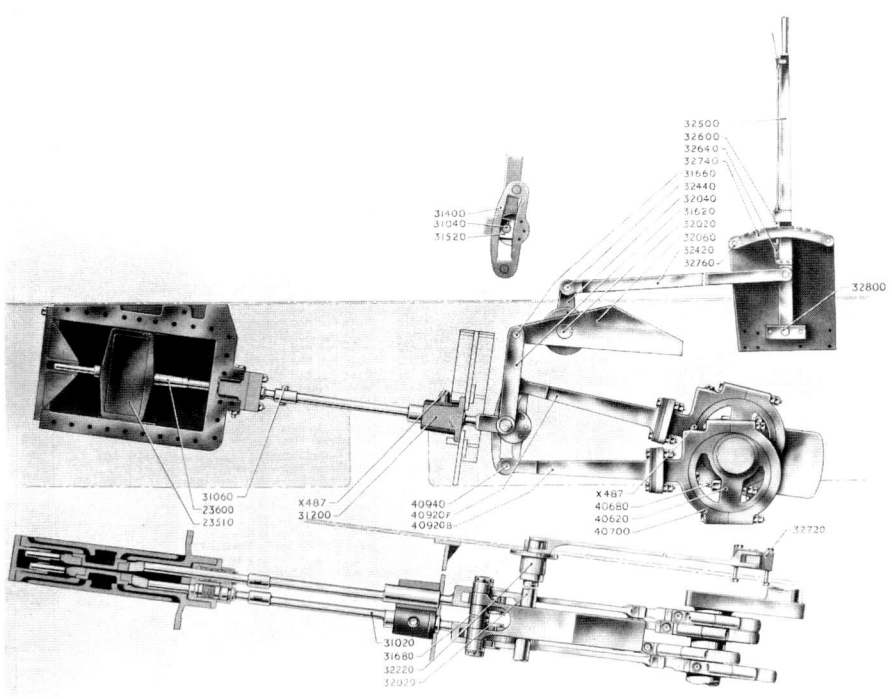

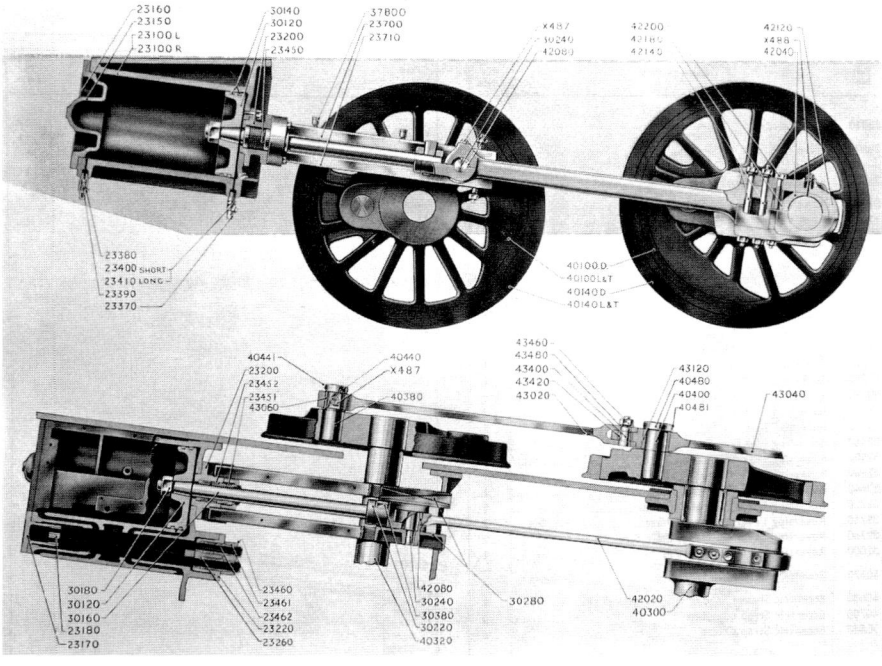

Joe Cunliffe, former manager at Walkden Yard, writing in 1990 (d. 1993) had only one criticism of the Austerity tank locomotives: that they were a bit top-heavy. This had contributed to three of them turning over: two at Ashton's Field Colliery and one at the very large Cutacre waste tip (south of the A6 at Little Hulton, opposite Brackley Colliery, opencasted and landscaped by early 2013). At Ashton's Field a British Railways breakdown crane was borrowed to right the engines; it came from Brindle Heath, the extensive sidings close to Agecroft Colliery (Salford) locomotive shed where a breakdown train was permanently kept. Joe recalled the track at Cutacre tip not being suitable for a heavy crane and the Walkden men had a difficult job.

On 3 April 1946, a Lancaster bomber of 90 Squadron RAF carried out aerial photography runs over the north-west. This view has been rotated, placing the north-west at the top edge, where Walkden Yard, its mineral line and Ellesmere sidings can be seen. Crossing lower left is the LMSR Wigan to Manchester line and heading nearly vertical on the left is the LMSR Bolton to Manchester line.

Wagon Shortages

The large fleet of Manchester Collieries wagons, commandeered during the war for the national pool, gradually arrived back, many in a sorry state, from 1945 onwards. A deal was made with the Central Wagon Company's works at Ince, Wigan, who purchased the wagons and leased them back. By June 1946 as nationalisation approached, the number of wagons had reached an impressive 9,515.

At the end of the war seventeen main-line wagons in the national pool had been out of commission, some under repair, and nearly seventeen privately-owned wagons were out of commission. fifteen locomotives nationally had been awaiting repair. As a matter of urgency the Minister of Transport, A. J. Barnes, made a statement in Parliament in December 1946 – the matter of materials shortages for coal-wagon repair shops being raised, he stated that the various railway companies had not been prepared for the additional wartime workload. As a result the stock of wagons had deteriorated rapidly. He stated that upwards of 500,000 wagons needed replacing. The minister stated that initially he would be placing an order for 50,000 steel-bodied wagons of 16-ton capacity.

Post-Second World War, coal was still the prime source of energy in Britain and politicians had their eyes once more on protecting this vital industry, nationalisation in the back of their minds. Strangely, Manchester Collieries were not deterred by this being on the horizon and they surveyed their whole mining system in detail and planned for major expansion in mechanised post-war output. Mosley Common Colliery was to receive the most attention due to the huge coal reserves lying to the south, assessed at at least 145 million tons. Most of the major redevelopment schemes were to be completed by the NCB after total nationalisation of the coal industry in January 1947.

Public Ownership, 1947 Onwards: The National Coal Board

In 1942, landowners and royalty owners had been compensated to the tune of £66,450,000 when the government took control of the industry. The coal mines had been taken under government control previously, during the First World War. In 1942, the industry was placed under the control of the Coal Commission with regional boards, but the mining industry itself remained in private hands.

The General Election of 1945 brought in a Labour government. The Labour Party had long desired to nationalise the coal industry. A Bill to carry this out was read in the House of Commons in December 1945. Royal Assent was given the following year – the Coal Mines Nationalisation Act 1946 was passed, with nationalisation 'Vesting Day' taking place on 1 January 1947.

Vesting Day ceremonies took place around the Lancashire coalfields, the main one being at Gin Pit, Tyldesley, and its associated Miners Welfare Hall. Others took place at Mosley Common Colliery; Lyme Pit, Haydock; Sutton Manor Colliery, St Helens; Parsonage Colliery, Leigh; Chisnall Hall Colliery, Coppull; Ince Moss Colliery, Wigan, and later at Oldham and Burnley. The men also received an immediate, and no doubt welcome, one-day holiday under their new terms of employment!

Huge potential amounts of compensation for the coal companies were now on the table for the colliery companies. Two former senior area surveyors, one with the pre- and the other the post-nationalisation industry in Lancashire, recalled to me that at times compensation was awarded for unworked coal that companies had never intended to work – one of the men had prepared 'proposed working plans' for thin seams such as the Smith.

In 1947, Manchester Collieries received compensation to the tune of £3,346,200. Most of the former managers and officials walked straight into lucrative posts within the new National Coal Board. For the men on the coalface or working the locos on the surface nothing changed immediately. I have been told this repeatedly over the years by many men who worked in the industry at the time. In the first year of nationalisation the industry suffered

1,635 strikes or disputes leading to loss of coal production, over 300 more than the previous year.

There had been a coal shortage in the international economy between the end of the Second World War and 1957. Demand was high and in early 1947, the first year of nationalisation, freezing weather highlighted and intensified the coal shortage even more, producing a serious energy crisis. The government was forced to ration coal and many households were left without heating. Exports were temporarily suspended and, for the first time, the government considered actually importing coal.

To protect the country from future crisis, the government attempted to increase capacity and production in the industry. This proved difficult as the post-war shortage of manpower persisted. In 1950 and 1956, the National Coal Board strived to raise production targets until the mid-1960s. As the established deep mines were unable to meet demand, the government invested in shallow opencast mining during the 1950s.

Only three pits of the once-vast Bridgewater Collieries mining complex based around Walkden had survived nationalisation: Brackley Colliery, Little Hulton; Mosley Common Colliery, Boothstown; and Sandhole (Bridgewater) Colliery. east of Walkden.

The following table shows the extent of Manchester Collieries and its workforce on nationalisation, January 1947:

Colliery	Location	Underground	Surface
Astley Green	S Astley	1,375	561
Nook	S Tyldesley	1,365	355
Gin Pit	S Tyldesley	362	158
Bedford	E Leigh	704	252
Chanters	SE Atherton	945	429
Gibfield	NW Atherton	530	178
Howe Bridge	SW Atherton	312	136
Mosley Common	E Tyldesley	1831	627
Newtown	Clifton	570	240

Brackley	E Little Hulton	761	271
Sandhole	E Walkden	725	335
Wheatsheaf	Pendlebury	659	274
Bradford	E Manchester	690	223

The system of using alloy pay tallies, in use in the Manchester Collieries era, continued into NCB days. (Charles Birdsall collection)

The colliery location maps on pp95-97 were included in *The Colliery Guardian Guide to the Coalfields*, in the first edition of 1947:

Nationalisation had created an industry that had 1,500 steam locomotives at work in 540 separate locations, previously owned by 260 private companies. Forty locomotive manufacturers were in operation, and a small number of locomotives was built by coal owners, Wigan Coal & Iron Co. Ltd for example. The coalfields were divided into forty-eight areas. A standard blue livery had been advocated but not always applied, along with locomotives occasionally being identically numbered at the same colliery, as renumbering and transfers took place.

In January 1947, Walkden Yard became part of the new No. 1 Manchester Area of the National Coal Board as its central workshops, the area comprising twenty collieries. Working conditions did not really change much but occasionally tensions arose, for instance when men working the Walkden section of the mineral-line system found that colleagues at Astley Green Colliery and Nook Colliery were being paid hourly rather than by the shift, leading to an overtime ban.

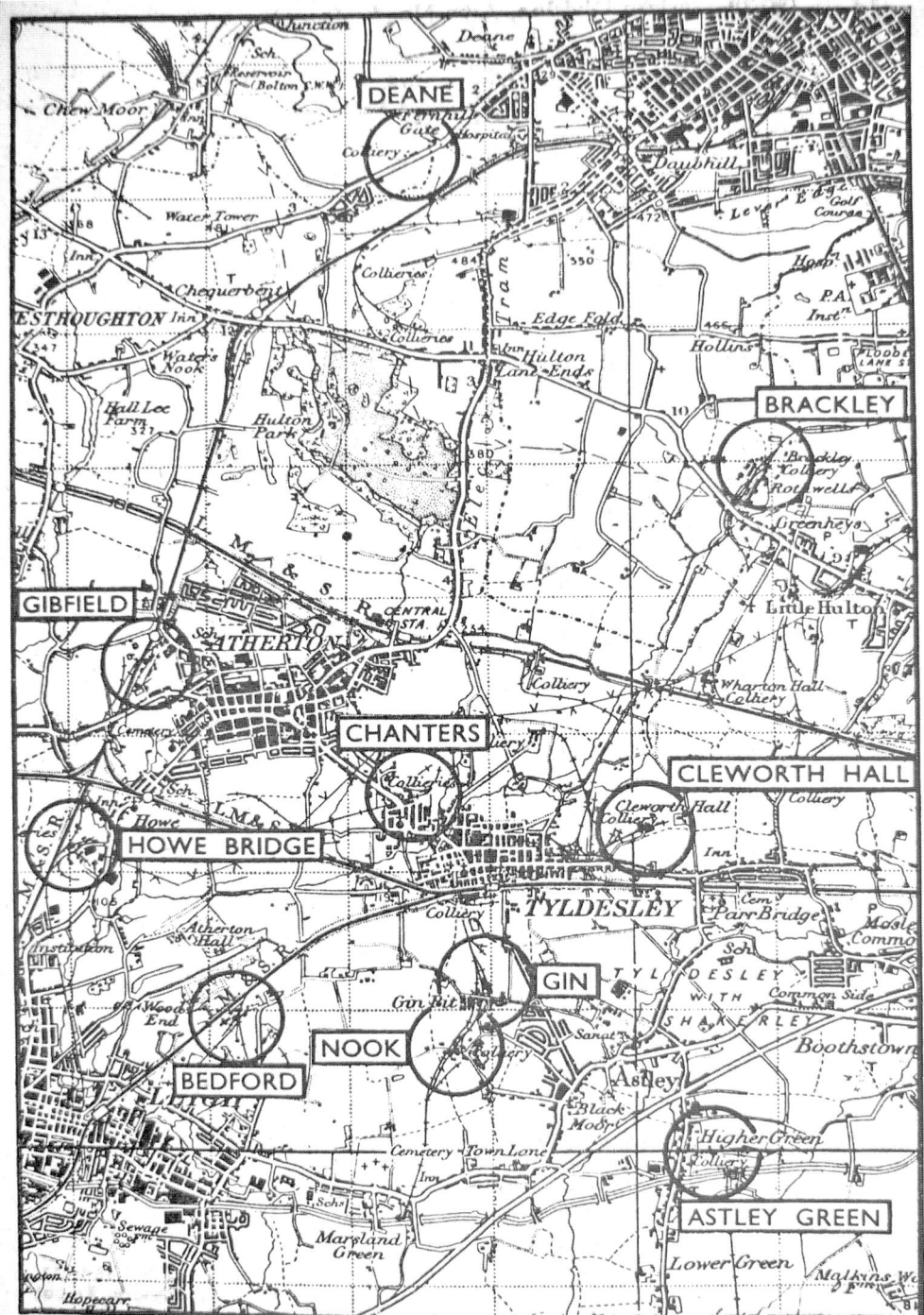

Based upon the Ordnance Survey Map with the sanction of the Controller of H.M. Stationery Office.

BRACKLEY
MOSLEY COMMON
SANDHOLE
Brackley Colliery
Rothwells
Greenheys
Little Hulton
Wharton Hall Colliery
Peel Hall
Makants Colliery
New Bury
Stoneclough
Ringley
Kearsley
Colliery
Hilton
Walkden
Linnyshaw Colliery
Elec. Tram
Wardley
Bridgewater Colliery
Parr Fold
Parr Bridge
Mosley Common
Ellenbrook
Roe Green
Hazelhurst
Common Side

NEWTOWN
SANDHOLE
WHEATSHEAF
Walkden
Linnyshaw Colliery
Clifton
Canal
Elec. Tram
Bridgewater Colliery
PENDLEBURY
Colliery
SWINTON
Roe Green
Hazelhurst
Moorside

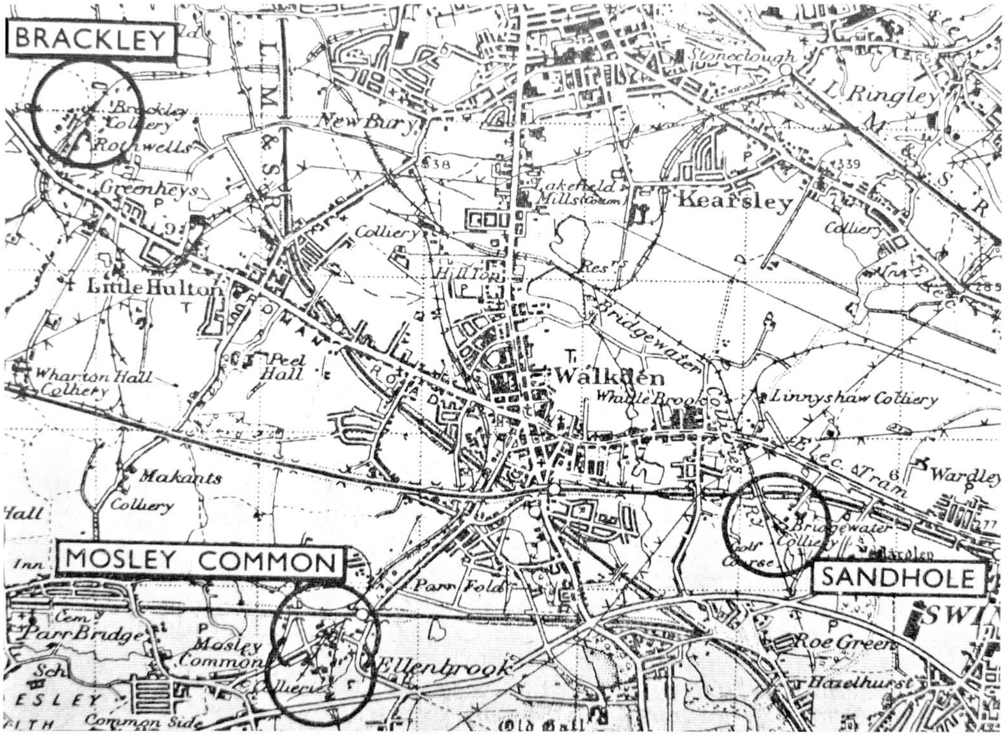

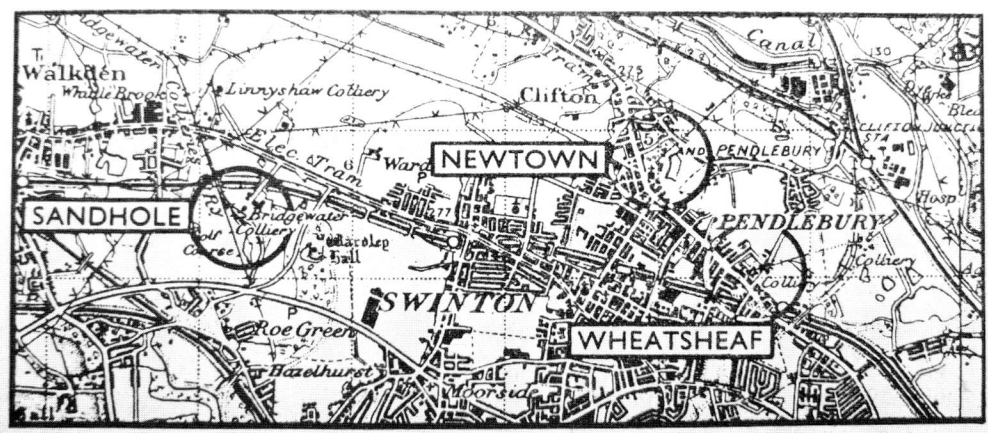

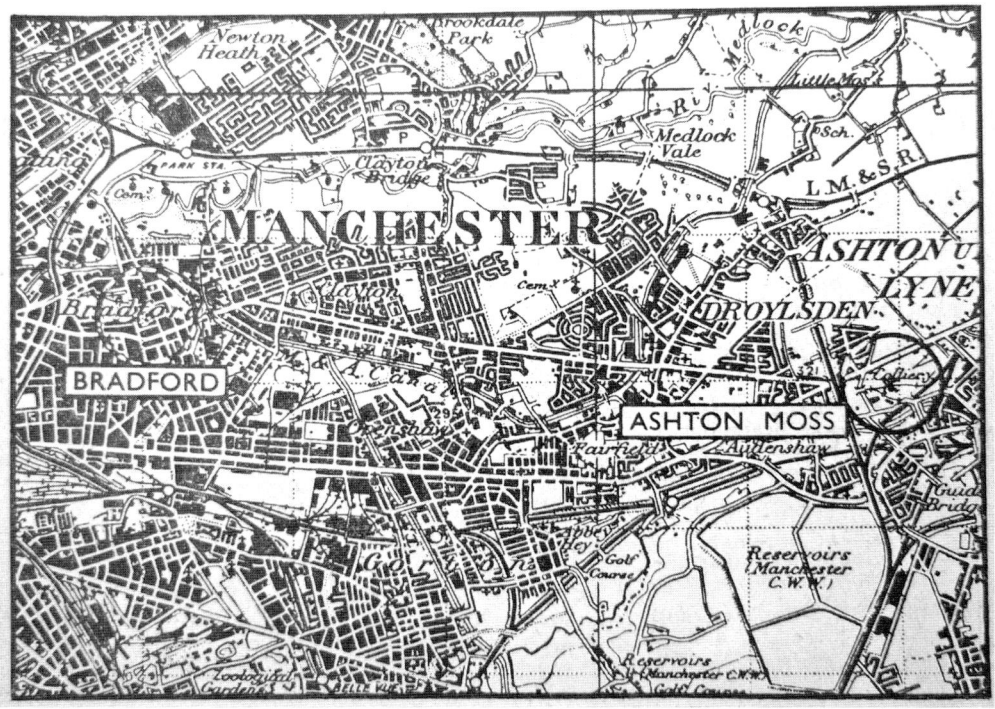

Walkden area locomotive staff pay inequalities were highlighted in *The Manchester Guardian* of 3 January 1951.

DRIVERS LIFT PIT OVERTIME BAN

Talks to be Reopened

Some forty locomotive drivers and brakesmen at Mosley Common, Sandhole, and Brackley collieries in the Walkden area of Lancashire, who for a fortnight have been refusing to work overtime or during the Saturday shift, decided by a small majority last night to lift their ban. The men assemble wagons of coal at the collieries and move them to the railways, and their action is stated to have reduced deliveries by 10,000 tons a week.

They are asking for payment on an hourly instead of a shift basis, which would increase their wages when working during a Saturday shift by 8s. 1d. They often work alongside men from the neighbouring Astley Green and Nook collieries, who are paid on an hourly basis and consequently earn more, and this has caused dissatisfaction.

Negotiations to end this anomaly were in progress, but were broken off by the North-Western Division of the Coal Board when the ban was imposed. They will now be reopened.

On nationalisation in 1947 only 2.4 per cent of coal nationally had been power loaded at the coalface using coal cutters and conveyors. By 1957 this had risen to 23 per cent and by 1972 it was 97.7 per cent. Walkden Yard could never have been expected to take on the engineering and maintenance work relating to twenty increasingly mechanised collieries. From January 1952 onwards Bickershaw Colliery, Leigh; Parsonage Colliery, Leigh; Bedford Colliery, Leigh; Howe Bridge Colliery, Atherton; Chanters Colliery, Atherton and Gibfield Colliery, Atherton, were transferred to the No. 2 Wigan Area. The old Wigan Coal & Iron Company workshops at Kirkless, north of Ince, took over the maintenance of the locomotives from these pits.

With apparently no fears either for the future of the coal industry or the continued use of coal-fired steam locomotives, a new locomotive workshop opened at Walkden Yard in August 1953, although the possible imminent closure of the Gin Pit, Tyldesley, workshops had been sensed.

The impressively spacious new loco maintenance facility that opened at Walkden Yard in 1953, photographed from the overhead crane. The building was slotted in between the Fabrication / Blacksmith's Shop and the Mechanics Shop. Lower left is the 0-6-0ST IC Robert Stephenson & Hawthorns WD 7174/1944 *WHR*, and to the right is the 0-6-0T OC Hunslet Engine Co. 1475/1924 *Bridgewater* in for repairs. The drums of steel winding rope in the distance were destined for Mosley Common Colliery.

A mid-1950s scene in the Erection and Blacksmiths Shop at Walkden Yard, a row of Massey pneumatic hammers in view. To the right a forge chimney hood can be seen.

A neat and tidy work area for these locomotive fitters in the shops at Walkden Yard around the early 1960s. The boiler is thought to have been from one of the 1920s 0-6-2T North Staffordshire Railway / LMS locomotives.

An impressive view of a loco boiler end-on in the Walkden Yard loco workshops. The boiler is one of the replacements supplied by Hunslet in 1945–9 and is thought by experts to belong possibly to 0-6-2T NSR / LMS (built Stoke 1923) *Princess* while being overhauled in 1960. It left Walkden on 22 April for Crewe Works for a repaint and then the Stoke-on-Trent railway exhibition as NSR No. 2. It returned to Walkden in June 1960 and was back in less glamorous service by August.

A fascinating and important Walkden Yard workshop scene of around December 1979, fascinating because all the locos in view are now preserved. The fitter is working on the valve chest and cylinders of probably Hunslet Austerity (3163/44, later HE 3885/64), ex-Bold Colliery *Alison* on 12 October 1979, which returned to Bold as *Joseph* on 20 February 1980. Today named *Sapper* and in action at the East Lancashire Railway. The second diesel loco on the left is a Bagnall o-6-oDM, probably *Wolstanton* No. 3 (WB 3150/60) at Walkden from about March 1979 until 7 August 1980. After Wolstanton Colliery closed it was bought for preservation and is currently on the Foxfield Railway. The diesel loco in front is a North British o-4-oDH and this would be D5 NB (27876/59) supplied new to Haydock, later Parsonage – Parkside – Parsonage, then to William Loco Shed at Whitehaven Harbour. To Walkden in April 1979 and returned to William in August 1980. This is also preserved, recently noted at the Chasewater Railway, Brownhills, Staffordshire. The fitter is working on the last steam loco overhaul completed at Walkden Yard.

At this time a typical workflow at Walkden and most central workshops could be:

  Job reception from colliery
  Stripping and degreasing
  Inspection
  Spares and materials assembly
  Workshops
  Inspection and testing
  Painting
  Return item to colliery or put in store

Failsworth was the first locomotive to be dealt with in the new workshops on 3 August 1953.

Norman Glyn Recalls Working at Walkden Yard, Sandhole and Mosley Common Collieries in the 1950s

You mentioned that your father suggested you might get a job at Walkden Yard.

Yes. My father had worked at Walkden Yard as a boiler stoker from around the Second World War period; he had previously worked at the Nelson Pit, Tyldesley [Wm Ramsden & Sons joined Manchester Collieries in 1935], doing the same work. He retired around the late 1970s.

  I was sixteen when I went to Walkden Yard in 1950 to see if any jobs were going. I waited on the wrong side of the manager's house for a couple of days – the Tynesbank side of Fred Hilton's house – until I realised he left the house by the back door, down the steps and into the yard!

  My dad said, 'I cawn't see eh they'd a missed him!'

So you eventually met Fred Hilton round the back?

Yes, just for a minute or so. I mentioned my dad worked there and he said, 'Right, come on Monday morning and see Mr Spencer, the workshop foreman'. There were two shifts then at Walkden Yard, days and nights, days was seven till four, nights ten till seven, there wasn't an afternoon shift. So I started and was led to the plating or fabrication [erection] shop, through some curtains behind which the welders worked, and pointed to a man who would 'show me the ropes'. I finished up with a guy called Ken Mather; he was a welder who worked on the locomotive boilers. Any boiler work and he was on it. Larger jobs like split or damaged locomotive frames usually had new sections made rather than repairing them by welding.

I gradually built up my skills, mainly at welding, until allowed to carry on work which the previous shift had started without supervision.

Boiler re-tubing seems to have been a regular job looking at the old maintenance records; why was this? What work was involved?

This was due to hot, often corrosive, water circulating around them and corrosive gases passing through them leading to pitting, general corrosion, cracking, scale deposits and other defects. One of the biggest jobs we did was welding up the smokebox end-plate bottom which corroded badly on all locos; that was an awkward job. We only had compressed-air grinding tools to smooth off the finish, nowhere near as powerful as today's grinders. Then we would leave it for the boilersmiths to ream the end holes out and fit new copper tubes with tapered ferrules on the end.

Ken Mather often did the finished weld as he was skilled and could leave a smooth weld finish which didn't need much grinding. I never really matched his skill.

Did you work on particular types of locos or all types?

We worked on any which came in; they all suffered with corrosion at the base of the smokebox. To be honest we just worked on whichever locos came in; we didn't really take notice half of the time which one or type it was. When I wasn't working on locos I would be working in the plating section, welding up for them.

What sort of jobs might you be working on in the plating section?

All sorts. If a cage came in [from a pit] smashed or needing repair it had to be turned round within twenty-four hours and back in use; sometimes work like that came in at the end of a shift – you could go home if you wanted, but often men would stay and get to work on it straight away, overtime!

What proportion of your work was repairing compared to fabrication?

I would reckon about 50/50. All sorts of mining equipment would come in: coal cutters, haulage engines, pumps, motors, the early Anderton disc shearers – they [shearers] were in their infancy then in the early 1950s, don't forget.

At the bottom of the shop we also had work going on with suspension chains, cages and butterflies [Edward Ormerod safety detaching hooks, which would separate the cage from the winding rope if overwound, invented in

1868 and still made at Atherton]. They had their own furnace for annealing detaching hook parts after crack detection testing. All the suspension chains were checked as well.

Colliery winding-rope samples were checked in a separate bay – they would split the section of winding rope into individual strands and check them. One unusual job I occasionally did was welding steel plates on the front end of coal barges at the Worsley boatyard [part of the original Worsley Yard]. I remember the big steamer for softening barge timbers; the wood was like liquorice sticks afterwards, you could have bent it into any shape.

What was the atmosphere like at Walkden Yard? Was it old-fashioned?

Oh yes, some of the older supervisors who had been there a long time had to be called 'mester' and wore suits with ties, pocket watch and chain. A lot of the equipment was really antiquated – steam hammers were still being used.

Did you ever call into the laboratory at Walkden Yard? What happened there?

Yes, that was on the east side of the yard on the other side of the railway lines, near to Bridgewater Offices. They carried out coal sampling; a clever bloke was in charge, Professor somebody. The men there had white coats on!

They had a small coal elevator there that occasionally needed welding. Normally the yard steam crane would lift the welding set across the fence for us but it was broken down so we had to push the welding set all the way up Tynesbank and round the top. It was a very heavy Metro Vick [Metropolitan Vickers] set with cast-iron wheels.

Any memorable occasions?

Not really, although I remember one day when all of a sudden everything went quiet and I couldn't figure out why. The supervisor told me to take a break as King George VI had just died [6 February 1952] so we all stopped work for a while out of respect.

Did you get to see most of the work that was ever carried out on locos?

Yes, I remember seeing the big lathe turning and truing up wheels – the journals used to wear, we used to build them up with weld in very small amounts so they could be turned on the lathe. Bearings were made of phosphor bronze. The wheel tyres were usually shot at. As soon as a loco came in for overhaul, the first thing done was to get the wheels off and re-tyre and true them up.

Do you remember the diesels arriving at Walkden?

Yes, around the mid-1950s. I was the only person who would do any welding on their tank drains, everyone else said they might explode!

Why did you finish at Walkden Yard in 1960?

I heard there might be one or two redundancies coming up – probably due to pit closures, the reduced amount of work, etc., so I decided, being one of the younger men there, to go for it and spread my wings. I went to work as a welder at Sandhole [Bridgewater] Colliery, based on the surface but occasionally going below ground to help with measuring and setting jobs out and erecting them, plus later on some actual welding on the track joints that the electric trolley loco ran on. Strict times were laid down when we could weld below ground – there would be three welders on the job. The Inspector of Mines was usually present, the ventilation current was gale force and men with fire hoses were dowsing any sparks as we began welding, so we ended up soaked!

At Sandhole I had to do the basic four weeks' training for anyone working below ground at a pit. I was taken to see a coalface at Brackley Colliery – it was low and the colliers cleared some dirt away so we could crawl on; not for me that type of job!

During the holidays, men from Walkden Yard would often come and work at the pits on jobs like bunkers or other big jobs that would have interrupted coal production. Men at Walkden were often on call anyway to go out to pits at short notice to carry out welding and other jobs. I remember doing some welding work at Outwood [Radcliffe] washery.

Was welding in coal washeries dangerous what with the possibility of fires or explosions due to the dust?

It could be if the dust was dry; we had a few fires. Once an inspection lamp fell and broke during the night at Mosley Common Colliery washery and set fire to the dust – it was well away.

Why did you leave Sandhole Colliery?

Well I heard that it might be closing in a few months [closed September 1962] and was told a job was going at Mosley Common Colliery so I took the job there. I was there for about eighteen months until I got fed up with the industry and eventually went to work at a welding and fabrication business in Bolton called Angle Bank. The owner later wanted to pack in so I bought it with a mate Ronnie who worked there; we then named it A&B Welding.

Walkden Yard motor body repair men were highly skilled in sheet metal fabrication. In this 1954 image, we have back row, left to right: Frank?, Eric Cheetham, Jim Shaw, Ted Evans, Jack Fowler, Harold Mort, Jack Williams. Front row: Les Woodcock, Albert Butler, Alan Goodier, Ken Smith, Ernie Freeman, Joe Cunliffe? (Gordon Rawlinson, Jack de Guillaume)

We both retired around 1998. Today, Norman continues to breed his award-winning Boston Terriers and uses his Walkden Yard skills making replica parts for Velocette classic motorcycles.

Cyril Golding Calls at Walkden in 1954

Railway enthusiast Cyril Golding recalled a chance visit to Walkden Yard on a quiet Sunday afternoon in April 1954 in an article first written for The Industrial Railway Society in 2003. Staff allowed him and his friend to potter round the site taking photographs. The remains of tank loco *Queen Elizabeth* (formerly North Staffordshire Railway, later ex-London Midland Scottish Railway 2270, 10/1937) stood close to the banking where Ellesmere Colliery pit headgear stood. Nameplates had been removed along with many other parts.

Deciding to explore the site, crew on board loco *Warspite* (Hunslet Engine Co. Austerity 3778/1952) welcomed them. Close by, the engine shed contained three more Hunslet productions: locos *Respite* (3696/1950), *Revenge* (3699/1950) and *Wasp* (388/1954). Also in the shed stood loco *Charles* (Hudswell Clarke 1778/1944).

The loco crew directed Cyril and his friend to yard manager Joe Cunliffe with a view to seeing the workshops. Joe gave them permission; the men walked round alone, Cliff noting a large, brand-new, green 0-6-0 500-hp Hunslet diesel, numbered 4003, being prepared for trials by Hunslet staff.

Cyril recorded the painting details for NCB NW Division Manchester Area 1 locomotives during 1954–56:

> Black, with single yellow line. Red buffer beams, rods and outside motion. No indication of ownership. Lining out limited to the following areas above the running plate, platform angles not lined;
> Saddle tanks: sides and front face
> Side tanks: sides only
> Cab: sides only
> Bunker: sides and rear face
> Splashers and boxes: outside face only

Cyril mentions that in 1957 a cost-containment exercise at Walkden Yard had greatly reduced the number of loco repaints, even after major overhauls.

Walkden Workshops Manager, Harry Joyce, retires in 1954 after nearly fifty years' service

In October 1954, Harry Joyce retired from his position as workshops manager. Having served his apprenticeship at the Cambrian Railway Company's extensive workshops at Oswestry, Shropshire, he had arrived at Walkden in 1906. After a spell at Ashton's Field Colliery he moved to Canada until the end of the First World War and returned to the colliery in 1919, serving there for eleven years before returning to Walkden Yard for twenty-four years until retirement. In an article in *The Farnworth & Worsley Journal* of 8 October 1954, Harry recalled that his first job at Ashton's Field had been making coal cutters, with the designs adapted from American machinery. Later during the First World War, supplies of American mining equipment dried up and Bridgewater Collieries embarked on production of their own design with Harry in charge of engineering operations.

He recalled that after the creation of Manchester Collieries in 1929 the pit-tub building shop at Walkden Yard was initially very busy (it was later transferred to Ashton's Field). Production of coal cutters ceased as a range of specialist British companies were now providing them.

The piece ended with Harry stating that 500 men were employed in 1954 at Walkden Yard along with 160 at Ashton's Field, mainly on pit-tub fabrication, and that canal boats were still being repaired at Worsley boat yard.

Forty-Two Years on the locomotives: Derek Lee looks back

Central Railways locomen Albert Longstaff and Derek Lee on board *Warrior* on 12 April 1969, the run being from Astley Green Colliery to Ashtons Field Colliery, the return loco only. (Steve Leyland)

In a letter of 30 July 1990 to railway historian and photographer Alex Appleton, Derek Lee, former Walkden driver, recalled his experiences working on the locos from 1942 to 1984. Derek worked for Manchester Collieries from 1942 until formation of the NCB in 1947, finally being made redundant in 1984. He progressed from fireman to brakeman to driver, his regular engine as a driver being *Charles*. Derek recalled that the brakeman was senior to the fireman, having a more complex and responsible job with aspects of safety involved. Ashton's Field Colliery was the hub of the system. The sorting of wagons with different grades of coal and sorting the empties on the various sidings were complex operations. Quite a lot of work was done at night in the heyday of the system.

Train movements were controlled by telephone from the weighbridges. Locations from which movements were controlled were, Derek recalled: Boothsbank, Ellenbrook, Walkden, Ashton's Field and Sandersons Sidings. In the event of movements that could conflict, instructions were given to, for example, put a train in a loop or siding to await the passing of another train. Derek said that the system worked well and he knew of no accidents caused by it.

Linnyshaw Moss, 21 October 1967. The line straight ahead led to Sandhole Colliery, the left-hand line to BR exchange sidings. Blackleach Country Park today includes the trackbed routes. (Philip Hindley)

When all the collieries were in full production, Walkden loco men worked three shifts: 6 a.m.–2 p.m.; 2 p.m.–10 p.m. and 10 p.m.–6 a.m. Usually Walkden locomotives did not operate beyond Linnyshaw Moss, but if traffic was particularly heavy elsewhere a Walkden engine and crew could be sent to assist. Derek gave an example that if Sandhole Colliery was very busy and a long train of empty wagons arrived at Sandersons Sidings, a Walkden engine would be sent round. Astley Green locomotives normally worked as far as the tip at Boothsbank (Boothstown Basin) and Sandhole Colliery locomotives worked as far as Ashton's Field Colliery. At Brackley Colliery, Little Hulton, one engine worked as far as Linnyshaw Moss and the other worked yard and Cutacre tip traffic.

Cutacre tip received waste from other collieries, for example Mosley Common, and the Brackley engine could be kept very busy. Derek emphasised the volume of coal going out at Linnyshaw Moss for LEP (Lancashire Electric Power Co., Kearsley Power Station).

Derek was asked about the merits of the Giesl ejectors and mechanical stokers. He made no comment about them other than that they were valuable as aids to reducing smoke emission. He didn't seem to think much of the various schemes to combat smoke, with one exception, and spoke in disparaging terms about the secondary air system. The most successful, he said, was the introduction of steam jets in the firebox, above the fire. The smoke nuisance was a serious business and the council stationed men on bridges over the line monitoring the smoke – they held something to their eyes [smoke density evaluation filters] in the manner of looking through binoculars. Drivers kept a lookout for them and passed word on that they were on the prowl.

In his earlier days Derek worked as the fireman on *Ellesmere*, which had been fitted with a new saddle tank of rather thin steel plate, reverberating in an annoying fashion as a result. There was a wagon turntable at Worsley [canal coal wagon tip] so that the end doors faced the right way. A driver was killed at Worsley when he was behind his stationary engine and some wagons being run down by gravity pushed the engine over him.

When there was no more work for him as a driver, Derek was transferred to the workshops as a mate to the fitters and boilermakers travelling widely in Lancashire and occasionally also going to the Cumberland and North Wales coalfields. He was very critical of the situation in Cumberland [the dreaded Ladysmith washery, the final repose of many colliery locos] and in particular the lack of equipment there. Derek stated that if you looked down the line towards Boothsbank canal tip with your back to Mosley Common Colliery, the sidings on the right were known as the Coronation Sidings, having been installed around the time of the queen's Coronation in 1953. Finally Derek recalled that firemen liked to get hold of main-line coal shovels which were superior to those provided by the NCB; main-line footplate-crew caps were also highly prized if they could be obtained.

Mick Hughes Recalls Working the Ellenbrook Sidings, Mosley Common, 1956–1960

As well as Walkden Yard the various sidings at Astley Green Colliery, Mosley Common Colliery, Ellenbrook and Walkden had staff of their own, sorting and shunting wagons of coal and occasionally dirt. I interviewed Mick Hughes in November 2012, now living in Atherton, who recalled his time at Ellenbrook sidings;

How did you start working on the locos?

Just looking for a job in general, I went to Walkden Yard and was interviewed by Harry Tweedy who was in 1956 aged fifteen. My father worked down Mosley Common Colliery No. 2 Pit and he used to walk up and down the mineral railway line to and from work; maybe somehow he knew Harry Tweedy.

My first job was at Ellenbrook sidings with a chap called Tommy Ogden. He had always worked on the locos and was probably nearly sixty-five then.

My first job was filling the water bottle up at Ellenbrook Brickworks as we had no water, then I was receiving telephone calls from Mosley Common Colliery sidings, telling us an engine was coming up with a train behind it, and it's going up to Walkden or Ashton Fields.

So there were no signals in use on the mineral line?

I had to liaise with men at Walkden by telephone and tell them an engine was on its way to them and they would say, 'Right-oh line's clear, off you go' [the Bridgewater Collieries, Manchester Collieries and NCB mineral-line systems had never utilised signalling; telephones at strategic sites controlled single-line working].

How was the approach to and up Walkden Bank tackled? Would you go full blast through there?

Oh crikey yes, you needed to go full blast up there and the maximum you could carry was ten wagons, and if it was dirt it was eight. I used to go on the locos at dinner time with a bloke called Tommy Tatlock. He was the [transport] foreman at Mosley Common [Colliery] and we used to jump on the engine and off we'd go together doing bits of shunting.

Do you remember the name of the loco?

Oh, I remember them all, often it would be a different loco each day.

What was a typical day?

I'd get to work for 7 o'clock and the empties came down at 8 o'clock – the engines came to us from Walkden Yard. They started at 7 o'clock but had an hour firing up, oiling and greasing, etc.

I remember Derek Lee [see Steve Leyland's photos] – he didn't have a regular loco, he wasn't a regular engine driver, he was the first spare one. He worked Astley [Green Colliery], Mosley Common [Colliery], Sandhole [Colliery] – all over the bloody place.

I didn't have a dinner, I used to go with Tommy Tatlock at dinner time to pick up a train, sometimes from Astley Green, sometimes Ashton's Field. We had a different loco every day. Once a week an engine used to do a double: 7 in the morning till they finished at night, officially 10 o'clock.

When I reached sixteen I graduated into becoming a spare fireman, mainly shunting in the Ellenbrook Mosley Common area. Now and again I was allowed to have a go at driving myself, oh there were no Health and Safety in them days!

All these different locos, did they have different characters? Different power?

They were all identical apart from the six-wheelers [the ex-North Staffs/LMS locos]: six plus two.

Were they more powerful?

They were more for high speed than for power – the little Austerity engines were like a bull at an entry, the six wheelers were more like a greyhound if you understand what I mean.

So if one of the North Staffordshire locos was doing the run to Ashton's Field could they manage more wagons?

No, exactly the same.

Did you ever do the run all the way to Ashton's Field?

Yes, I did go to Sandhole, but that was for running the coal off from Sandhole washeries on nights because they were short [staffed].

Any memorable occasions during your time on the locos? Accidents, etc.?

No, not really, don't remember any accidents. There had been one when I wasn't there – one of those big ones ran off the track but I don't know any details. I saw wagons come off the track, we had to get them back on ourselves using ramps, but ramps weren't carried on the loco, they were located somewhere else.

Was it tough in winter on the locos?

It was very nice and warm with the heat from the firebox.

Coming down from Ashton's Field past Walkden would someone have to jump off and pin the wagon's brakes down?

Oh, yeah, the locos' brakes wouldn't manage it, no chance, not a cat in hell's chance. When the last run from Ashton's Field was coming down with thirty or forty wagons, well just imagine what weight there was there. Maybe 30 per cent of the brakes would be pinned down.

Would the brakes get red hot?

Oh aye, they would get red hot and sometimes they would jam on [sieze] so they'd get flats [flat spots] on the wheels. A number of these at once would create a bump, bump, bump sound. I remember going on one [NSR/LMS] engine that had been dragged down the road; it had flats on all six wheels. I have a feeling it were him [pointing at a photo of a driver], because he were a crap driver; all I can remember is his horrible set of teeth. Some of the old drivers were bastards: you'd be shovelling away, doing your thing, then all of a sudden the bloody [fire] door would be shut – he'd kicked the bloody door shut. 'There's enough on there' he'd say.

Then there was another fellah, he was that old and decrepit he wasn't strong enough to bloody pull the regulator open, all the way I mean, so you had to get on the other side and give him a lift pulling it over!

I used to call in Walkden Yard occasionally. I remember Tweedy [Harry Tweedy, superintendent at Walkden Yard, loco *Harry* was named after him] wanting to be called Mr; he liked that.

Was there a team of men who took care of the line itself?

Yes, they were based at Ellenbrook Sidings. One man, Jim, had the biggest oil can I've ever seen in me life, for oiling the points every morning. He used to walk the full length from Mosley Common sidings, east, west and centre sidings, to Ellenbrook. I would say there was half a dozen gangers at the most, based at Ellenbrook.

[Mick didn't remember any stalls occurring on Walkden Bank, he suggested this might have occurred if the wagons had dirt in, extra weight. Dirt travelled in, as he termed it, 'white banders' for internal use only – usually old wooden-sided wagons whose brakes more often than not didn't work.]

Did you ever see any locos which were from 'out of the system', not related to the pits?

Once a week a loco came from the main line which took coal to Douglas (well, Liverpool) in the Isle of Man; that used to take about forty wagons of best coal. The locos were monsters, bloody giant things.

Did the diesels arrive while you worked at Ellenbrook?

No, the only one I saw worked at Sandhole [Colliery]. It had just arrived then; it wasn't up to much to be honest, I think they got stronger engines after that.

Did the crews take pride in their locos?

Oh yeah, every day the fireman would clean the front of the loco, brass work, etc., nameplate. The locos were very clean inside the cab, bloody spotless some of them. The men who used to light the loco fires in Mosley Common shed on nights used to clean one loco between them each night.

A typical day finished at 4 o'clock; we worked Saturday mornings as well from 7 till 12 – we got time and a half [overtime pay] for Saturday mornings, and I used to work one week of my holidays – you only got two weeks' holiday a year – for that I would get 29 shillings and four pence a day, whereas at Ellenbrook I would normally be on 16 shillings and eleven pence a day, so it was a big bloody difference!

What sticks in your mind about the period you worked on the locos? Did you enjoy it?

Oh it was bloody brilliant, I loved it, best job I ever had. I remember doing a 'double un' [double shift] where I ended up with four days' wages. On the last day of the working week I used to be allowed to jump off the loco and head home as I lived near the railway line at Mosley Common, so there was only a brakesman and driver taking the last load up to Ashton's Field.

Was there ever any training involved, local technical college for instance?

Oh, no, nothing like that, you just learned on the job – imagine, a fifteen-year-old lad in charge of an engine!

Being based next to Mosley Common Colliery did you ever wander into the pit buildings and chat to miners?

Oh yes, when we used to go for our wages or change our overalls on Fridays, but we didn't use the pithead baths at the colliery.

Was the coal separated by grades in the sidings at Ellenbrook?

Oh yes, 1 East siding was always full of shit, rubbish coal that had to go back through the washery system again; other sidings contained coal for specific customers. Coal reached the main lines at Ellenbrook, Walkden and Ashton's Field. Occasionally I would deliver dirt to Ashton's Field, which would then be sent to be tipped at Cutacre [south of Brackley Colliery, Little Hulton].

Why did you finish eventually?

Couldn't see a future in it, dead end job weren't it really? I went from there to be a bus driver with the LUT then a variety of jobs, including eventually offshore work in Saudi Arabia.

Today Mick keeps up his interest in local railway history and his passion for all things motorcycling. It is interesting that Mick comments that a different loco was in use virtually every day at Ellenbrook Sidings. Former Walkden Yard manager, Joe Cunliffe (died 1993), writing in 1990 stated that Walkden-based loco crews normally kept to one engine, with a crew comprising driver, fireman and brakesman. The fireman assisted the brakesman at, for example, Ashton's Field Colliery.

Further Expansion at Walkden, the Clean Air Act 1956 and Locomotive Modifications

The 1956 Clean Air Act introduced a number of measures to reduce air pollution after the deadly smogs of the early 1950s, creating smoke control areas in some towns and cities in which only smokeless fuels (anthracite or coke) could be burnt. Previous legislation dated back to the Public Health Act of 1936 where the term 'smoke nuisance' featured, mainly dealing with isolated occurrences and the abatement of such rather than legislation in force full-time.

The 1956 Act promoted alternative sources of heat such as the use of cleaner coals, electricity, oil and gas, the legislation reducing the amount of smoke pollution and sulphur dioxide from household fires. The Act placed certain obligations on the users of boilers where the regulations applied. For instance, the emission of dark smoke was prohibited except in specific circumstances, and practical measures had to be taken to prevent the discharge of grit and dust from boiler plant burning more than 1 ton of fuel per hour.

The section of the 1956 Act dealing with locomotives was actually not that harsh and was non-specific comprising only three short paragraphs, the most important clause being 'In addition, the owner of any railway engine must use any practicable means there may be to minimise the emission of smoke, and is guilty of an offence if he fails to do so and smoke is emitted. Nothing else in the Act apples to smoke, dirt or grit emitted by an engine'.

By coincidence the cost of suitable grades and qualities of coal for locomotive use around this time became lower, so the impetus was there for improvement in thermal efficiency and flexibility of operation of the colliery steam locomotive.

Long-standing manufacturers of locos to the mining industry, the Hunslet Engine Co., Leeds, rose to the challenge, applying design changes to a number of NCB 0-6-0 saddle tanks from 1961 to 1965. A mechanical underfeed stoker, devised by Thomas Hill (Rotherham) Ltd and powered by a steam engine, drove a trough and feeder assembly located under the cab floor. Coal was gravity fed to the trough, travelling along the stoker to the grate where the fuel spread giving a hot thick fire at the back and sides of the firebox. Ash gravitated towards the front to be emptied into the ash pan manually. Hunslet also introduced a 'gas producer' system in an attempt to avoid choking the grate with fused ash or clinker by cooling the ash with steam. As well as cooling the ash water, gas was produced – a mixture of hydrogen, carbon monoxide and carbon dioxide from the steam – adding to the combustion gas formed in the fuel bed. The steam came both from the mechanical stoker engine and the exhaust steam from the locomotive. The cooler fuel bed permitted the use of smaller coal, ½ in singles, thus reducing fuel costs. The locomotive driver controlled this additional steam to counter clinkering up.

Another modification to appear, easily recognisable in old photographs or restored locos from the associated chimney design, was the Giesl Ejector, a suction draught system invented in 1951 by Austrian engineer, Dr Adolph Giesl-Gieslingen. The Giesl ejector replaced the existing smokebox blastpipe with several small fan-shaped blast pipes, creating improved suction draught. Giesl claimed use of his ejector could lead to fuel savings of 6–12 per cent, in practice about 8 per cent. A power increase of up to 20 per cent was also

A Giesl ejector during testing at Rugby Locomotive Testing Station in 1959. An adapted version was to be fitted to many of the locos described in this book from the early 1960s onwards in an attempt to curb black smoke after the Clean Air Act of 1956.

The loco alongside Walkden Yard is possibly *Warrior* but the view interestingly shows the stark impact of the Giesl conversion on the locomotive chimney structure. (Roger Fielding)

claimed. The first Austerity locomotive to be tested was 2859 at Baddesley Colliery, Warwickshire, in 1959. In total, Hunslet and the NCB modified fifty locomotives, mostly in the Yorkshire and Lancashire coalfields, the last modification order being placed in 1968. Interestingly British Railways tested the system on just two of its locomotives and took it no further.

Overall the associated modifications had helped solve the problem of incomplete combustion with resultant thick black smoke, pollution and inefficiency. Local authority smoke-abatement officers were not to be convinced by such claims, and were to be seen at the trackside around Mosley Common and Walkden monitoring emissions with their smoke density Ringelmann charts held to the sky. Smoke density was assessed on a scale of 1 to 5 – 0 being white, 5 being pure black ('dark smoke'). A £100 fine was in place for 'dark smoke' producers. The loco crews must have had quite a struggle to comply with the new regulations as they battled up Boothsbank or Walkden Bank!

Walkden HQ

By the time the *Colliery Guardian* produced its series of coalfield maps in 1959, Walkden Yard and offices was the administrative and maintenance hub for the following collieries: Agecroft, Salford; Ashton Moss, East Manchester; Astley Green, South Tyldesley; Brackley, Little Hulton/Farnworth; Bradford, East Manchester; Cleworth Hall, Tyldesley; Deane, West Bolton; Mosley Common, East Tyldesley; Newtown, North Swinton, Nook, Tyldesley; Sandhole, East Walkden and Wheatsheaf, North Swinton/Pendlebury.

1961 saw further area reorganisation with the new East Lancashire Area being created and Walkden Yard taking over major overhaul work on all the locomotives in the North West Division, including North Wales. Locomotive work ceased at Kirkless, Wigan, in November 1962 followed by Haydock in March 1963. A 50-foot extension to the loco fitting shop at Walkden helped create space for the additional workload.

During this period locomotive enthusiasts were treated to the sight of long lines of colliery locomotives at Walkden Yard as collieries began to be rapidly closed by the NCB due to reduced demand for standard bituminous coal. During the period 1965–1970 the NCB had reduced the number of mines from 530 to under 300.

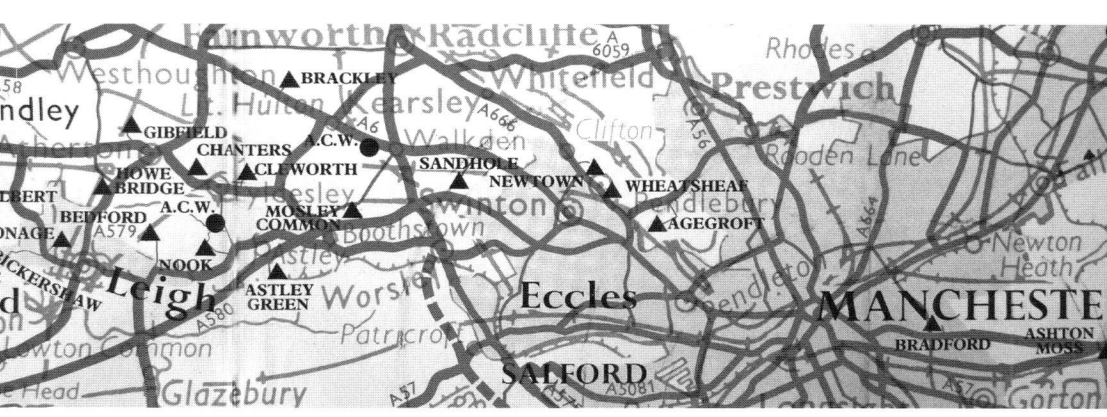

The *Colliery Guardian* published a series of folding coalfield maps in 1959, No. 5 being of the North West Division. This section shows the collieries across the coalfield, Walkden Yard being marked as A.C.W or Area Colliery Workshop.

Redundant loco line up at Walkden Yard, 27 March 1968. Right to left: *Atlas* from Bedford Colliery, Leigh; *Fairfield* from Ravenhead Colliery, St Helens; *Sir Robert*; *Renown*; *Wasp*; *WHR*; and *Bridgewater*; all were to be scrapped. (Philip Hindley)

Cliff Bray Looks Back at Three Generations in the Industry and Walkden Laboratories in the Sixties

My family had worked in the coal industry locally for the past three generations at least, employed in work ranging from sinking shafts and mining coal to surface work. My father (who died in 1960) worked initially in his younger days on the Central Railways for Manchester Collieries as a fireman (I have a photo of him on the footplate of 'Bridgewater') and subsequently (post-war) as a mobile crane driver in Walkden Yard – although the crane very infrequently went to help on day jobs at other collieries. He had a large allotment at the back of our house next to Tynesbank. He grew tomatoes, bedding plants, veg and chrysanthemums outdoors and in our heated greenhouse (made of surplus ex-Manchester Collieries/NCB stuff). A lot of the produce was sold to his mates in the yard. In the 1950s he also provided the bedding plants that were used to adorn the raised bank opposite the time office at the entrance to the yard – which always looked colourful and immaculate at that time. He died on 27 December 1960 of a heart attack in the belt shop at the yard – he was working on a bank holiday to clean out the boiler room in one of the workshops. His funeral was attended by a large number of colleagues who collected money to help his widow and family – typical of the friendship between workmates typifying the yard at the time.

My father, mother and I lived on Manchester Road East (A6) about 100 yards from the top of Tynesbank, so my father walked to work each day (or at lunch times sneaked back into the yard after lunch via his uncle's back garden opposite the end of Tynesbank, so avoiding having to pass the time office at the main gate). As long as I could remember, four o'clock in the afternoon Monday to Friday meant that hoards of blackened, boiler-suited men swamped the front and backs of our terraced house as they made their way home.

I worked in the laboratories adjacent to the yard (next to the drawing offices immediately over the footbridge next to the engine sheds) during the summers of 1962–1966 inclusive while at school and university. Mr Ledwards (the manager) who knew my mother and father helped me get the job. A Mr Kerley was the chief scientist at the labs at the time and he interviewed me before offering me a vacation job – which was then renewed annually! In the labs, we analysed coal samples from all over the area on a daily basis for moisture and ash content, calorific value, sulphur content, etc., performed mine air analysis, water samples and mine roadway samples in a work environment the like of which I've not seen since! Great characters spanning the generation gap, which was becoming wider as the 1960s progressed. The labs at Walkden Yard closed in 1967 and the work was transferred to labs in the Wigan area so I had to find work elsewhere!

Brian Wharmby (born in 1950) Recalls Walkden Yard Life

I lived in Walkden from birth (1950) until I left the area to go to university in 1969, emerging as an electrical engineer. I lived initially in Mayfield Avenue, then on High Street, both of which were only a stone's throw from Walkden Yard. An interesting feature of not having lived in Walkden for over forty years is that my mental images mainly relate back to those days, and haven't been diluted by more recent local changes.

Living in Mayfield Avenue meant that the influence of the NCB system was never far away. The sound of locomotives heading up from Mosley Common was ever-present, and my mum used to curse days when there was a strong westerly wind that used to blow coal smuts from locomotives on to her washing in the back yard. As children we had a great playground on the area behind the NCB offices on Bridgewater Road and facing Walkden Yard, known locally as the Office Field. In the early days, this was mainly grassy scrub but over the years there were developments.

Portakabin-type offices were built, together with a car park, and around 1960 a coal chute [elevator] was added at the North End to allow road lorries to tip coal into railway wagons below, but some parts remained untouched.

2-year-old Marion Wharmby on Office Field behind Manchester Collieries' Offices, Walkden, 3 May 1943. In the spirit of 'Make Do and Mend', she is wearing a coat made from her dad's old trousers. The field and its air raid shelters became an ideal playground. Ellesmere Colliery stands in the distance. (Brian Wharmby)

A particular attraction was the underground air-raid shelter at the northern end of the field. It was presumably built to house the occupants of the nearby offices during the war and had quite a complex structure. It was built beneath a large mound of earth with two entrances at ground level, one immediately behind the offices and the other at the opposite end facing the pithead gear of Ellesmere Colliery. These had quite elaborate brick arches. The internal tunnels in between didn't go in a straight line; there was at least one bend which made it very dark inside. At the centre, on top of the mound, there was a manhole cover covering a short shaft into the tunnels. This had an iron ladder. The mound and surrounding area, right up to the fence of the nearby nursery were covered in Japanese Knotweed. You can imagine that this was an ideal environment for children's games, hiding in the tunnels, and using the knotweed stalks as spears (I still have a piece of such a spear embedded in my thigh).

The substation next to the coal chute was a bit frightening to us youngsters, with its continuous hum. Behind this was an overhead cable duct that spanned the railway. One of my more fearless friends once climbed right across to the other side, but I was too much of a chicken to follow him. We rarely ventured on to the concrete footbridge that led to the engine shed, and the grounds of the defunct Ellesmere Colliery held few attractions.

But of course there were the trains. I remember all the locomotives, both operational and in store, that appeared in and around the shed from the mid-1950s until the demise of the system. I got my first camera in 1965 and took railway photos including the NCB system during my teenage years.

Competing Fuels and Decline

The coal industry was finding it very difficult to compete against cheap oil and gas. Collieries were closed by the NCB even with viable and substantial reserves in hand, the number in the North Western Area falling from forty-one in 1962 to twenty-one in 1967.

1962 saw the arrival of the new colour scheme for colliery locomotives – the rather gaudy maroon with yellow lining, with the National Coal Board crest on the side of the cab.

The hard-hit coal industry was losing its attraction to the school leaver viewing a possible lifetime career, yet the NCB continued to try and attract lads into the industry as miners or apprentice fitters and electricians. An open day was held in 1966 with various events on site: rescue displays, the work of canaries in mines rescue being explained, plus the main attraction for the children – the locos smartened up and accessible. The loco scrap line, *Jessie* prominent, was moved to the south out of immediate view!

1964 had seen changes at Walkden Yard as it became one of the divisional general engineering workshops covering the North Western Division of the NCB, servicing no less than thirty-six collieries and depots in Lancashire, Cumberland and North Wales. In 1967 the site took on the role of regional services with specific workloads: powered supports (hydraulic coalface roof supports), Gullick pumps (hydraulic), tanks and locomotives (underground). Additional work consisted of cables (mechanical and electrical), cage suspension gear (pit cage chains, shackles and detaching hook linkages), steam locomotives, lifting gear (pull lifts, tirfors, spreader chains etc), general fabrication and mechanical engineering. One-off engineering jobs were also called for from the collieries.

Wasp, *Harry*, *Sir Robert* and the mobile loco maintenance brake van line up for visitors at Walkden Yard's open day, 14 August 1966. An exciting display of empty cable reels adds to the thrills. (Brian Wharmby)

North Stafford 2 (formerly *Princess*) with access steps for visitors at the Walkden open day, 14 August 1966. The loco was getting used to being the centre of attention – it had been restored / renamed for the City of Stoke-on-Trent Golden Jubilee Celebrations of the Federation of the Six Towns in 1960. (Brian Wharmby)

Above: Walkden Yard open day, 14 August 1966. A dramatic rescue display for the visitors, this one a casualty probably of a heavy chip shop lunch and seemingly too late to be rescued. (Brian Wharmby)

Right: In 1966 the Area NCB HQ staff at Walkden offices generated this local newspaper advertisement trying to entice young lads into the declining industry, either as mining trainees or apprentice fitters and electricians. Collieries also had boards erected trying to interest men in 'A Job in Coal'. The open day in August also played its part.

COAL MINING
offers Apprenticeship & a Lifetime's Career

Boys entering coalmining today can look forward to a lifetime in the industry. The new streamlined Coal Industry will have fewer collieries but they will be big ones and highly mechanised. And there's an Apprenticeship waiting for every boy accepted.

GOOD PAID HOLIDAYS
2 weeks annual holiday, plus 6 statutory holidays, plus 7 rest days — all on full pay.

GENEROUS ALLOWANCES & BENEFITS
Sick pay, assisted travel, cheap coal for householders, all course fees paid. Canteen and pithead showers, club and sports facilities.

GOOD PAY right from the day you start.
These are the rates of pay you are paid during apprenticeship.

send this coupon for
FREE BOOKLET
To the Manpower Officer,
National Coal Board,
Bridgewater Road,
WALKDEN

Please send me the free booklet 'MINING APPRENTICE'

Name

Address

AGE	SURFACE	UNDERGROUND
15	£6. 3.6.	£6.17.6.
15½	£6. 7.0.	£7. 0.6.
16	£6. 9.6.	£7. 3.0.
16½	£6.12.6.	£7. 6.6.
17	£6.16.0.	£7.10.6.
17½	£6.19.6.	£7.16.6.
18	£9. 3.0.	£9.19.6.

The Late 1960s, Steve Leyland's Important Record

Coalfield reorganisation in January 1961 had split the North Western Division of the NCB into West and East areas. Walkden Yard was now in the NCB North Western Division, East Lancashire Area. The headquarters of the new area was also at Walkden at the old Bridgewater Trustees (and later Manchester Collieries) offices nearby. Sandhole Colliery (the former Bridgewater Colliery) closed in September 1962 leaving 1,043 men looking for a move to the remaining pits. As it closed, the eastern arm of the Central Railways network south of Sandersons Sidings was abandoned. Further coalfield reorganisation in 1967 created the NCB North Western Area (surviving until 1974 when the NCB Western Area was formed, with headquarters at Stoke-on-Trent). In 1968 the Clean Air Act was revised, mainly impacting industry with its demands for higher chimneys – no new pressure was to be laid upon the humble colliery shunting loco.

In *Industrial Railway Record* (Journal of the Industrial Railway Society) No. 172 of March 2003, Bolton resident Steve Leyland wrote an article along with his photographs of the final two years of the NCB Walkden internal railway. This fine piece of research and documentation will prove to be an important historical record for future generations. The society and Steve have kindly allowed me to quote from the piece and use a small number of Steve's photographs.

Steve began to visit the Walkden-based system regularly from around August 1968. By now the line which had once served a number of collieries was basically a run from the still-producing Astley Green Colliery, to the south-west of Walkden Yard, past Walkden Yard and on to Ashton's Field coal blending, sidings and disposal point to the north of Little Hulton. Here the link via Linnyshaw Moss sidings to the BR branch line and Kearsley Power Station could be made. At Astley Green Colliery a line headed south across the Leigh branch of the Bridgewater Canal to the colliery waste tips.

Steve Leyland made around sixty visits to the Walkden system until final closure in August 1970 as Astley Green Colliery itself closed. He reckons he travelled approximately 450 miles on the footplate with the generous crews who probably realised the timely importance of his work [but would probably never admit to it] and lucky for us the Health & Safety At Work Act was four years away.

Steve recalls the men speaking of the locos as '*Loco Fred*' to avoid confusion with men of the same name. The seven locos in operation from August 1968 were:

Warrior	(Hunslet Engine Co. 3823/1954)
Stanley	(Hunslet Engine Co. 3302/1945)
Respite	(Hunslet Engine Co. 3696/1950)
Harry	(Hudswell Clarke 1776/1944)
Witch	(Hunslet Engine Co. 3842/1956)
Humphrey	(Robert Stephenson & Hawthorns 7293/1945)
Repulse	(Hunslet Engine Co. 3698/1950)

The last four were usually based at Astley Green Colliery. *Respite* was spare, occasionally working at Astley Green Colliery. Steve recalls *Humphrey* and *Witch* being withdrawn from service by September 1968. Walkden Yard had two drivers, Albert Longstaff and Bill Unsworth. Steve only recalls being with one fireman, Derek Lee, and one brakesman, Ernie Gregory. Astley Green had more crews, with 'Mad' Bobby and Kenny Whittle being memorable.

The Great Triumvirate of locos *Harry* (HC 1776/1944), *Warrior* (HE 3823/1954) and *Stanley* (HE 3302/1945) ready for action at Astley Green Colliery, 4 April 1970. The photo was taken the day after coal production ceased at the pit. (Steve Leyland)

Respite (HE 3696/1950) at Astley Green Colliery, 10 May 1969, with brakesman Ernie Gregory on top. (Steve Leyland)

Steve described the run from Astley Green Colliery as follows:

> Power station coal was the staple commodity of Astley Green at this time
> and that bound for Kearsley made the longest trip to British Railways
> via Ashtons Field. The first 1.1 miles from the pit, straight and virtually
> level, mostly ran parallel with the Bridgewater Canal [Leigh Branch]. At
> the taxing left hand curve of Booths Bank, the railway crossed the 100
> feet contour and trailed into the original line, which went straight down
> to Boothstown (or Booths Bank) canal basin and wagon tip. Facing more
> or less north for the next 2.8 miles, the adverse gradient averaged 1 in 54
> until the 300 feet contour was reached near the top of Walkden Bank.

As with many NCB lines the severity varied greatly within that distance, but
quite a lot was at about 1 in 30. The route through Boothstown, under [firstly
Leigh Road then] the East Lancs Road, by the closed Mosley Common Colliery
[closed February 1968 with the loss of 2,701 jobs] and under the former LNWR
Leigh loop line (still open until May 1969) to north of Ellenbrook demanded
less arduous work from the locomotives. The ex Lancashire & Yorkshire
Walkden High Level line loomed ahead on its embankment and bridges.

It was now tough going as far as the [Walkden Yard] workshops, before
a brief respite preceded the formidable Walkden Bank itself, 0.7 of a mile in
length, immediately after trains had plunged under the A6 road bridge just
west of the town centre.

Repulse heads up Boothsbank towards the Leigh Road bridge, Coronation Sidings and Mosley Common Colliery ahead, 25 January 1969. (Steve Leyland)

In this image taken on 31 January 1970, *Repulse* is banked by *Warrior* with no less than forty-two wagons. The closed Mosley Common Colliery stands behind. The run is from Astley Green Colliery to Walkden Sidings and the total weight is nearly 800 tons plus loco weights. (Steve Leyland)

A view of Boothsbank (Coronation) Sidings looking south-east on 24 May 1969. The East Lancs Road is behind the unknown photographer and the loco is on the line which passes below Leigh Road towards Boothsbank tip and Astley Green Colliery.

Warrior (HE 3823/1954) with nine wagons on storms up Walkden Bank at 09.05 hrs on 4 April 1970. (Steve Leyland)

Warrior (HE 3823/1954) banked by *Repulse* is seen around 1969 at the rather basic (in modern terms) Grosvenor Street level crossing at the top of Walkden Bank. Typical ex Bridgewater Collieries / NCB trackside power pylons can be seen in the distance. (Steve Leyland)

This photo was taken on board *Harry* (HC 1776/1944), approaching Ashtons Field coal blending plant before its reopening on 10 May 1969. (Steve Leyland)

A kink to the right, ⅓ way up, marked the start of the marginally steepest pitch of 1 in 18, according to my estimate (officially 1 in 24). By Ashton's Field the distance from Astley Green was 3.9 miles, and the journey had taken around 22–25 minutes, depending on the circumstances.

The true enthusiasm of Steve and his friends can be seen when he recalls a typical 'meet':

> From our early visits, the Walkden men became used to picking up my friend and I close to, or at the engine shed, soon after 7.30. We often rode with the men most of the day when up to 35 miles could be covered, but no two days were alike. The amount of traffic for Kearsley varied, but it usually amounted to between 40 and 80 of the standard 16 ton wagons making the Astley Green – Ashtons Field – Lynnshaw Moss journey.
>
> Often we started off with a 'banker load'. All the banking was carried out by Astley locomotives and men, but sometimes they would take up a single or even banker load on their own. The restricting factor on through trains to Ashtons Field was, of course, Walkden Bank. The general limit was ten wagons per locomotive, meaning that twenty could be taken if a banking engine was employed. Sometimes an extra wagon for each locomotive would be attempted. Experience supported the Walkden men's widely held view that their opposite number used to work the banking locomotives too hard, too early, and end up short of steam on the most grinding final stretch, or even earlier.
>
> The Walkden men handled their locomotives intelligently, judging each stage of the long climb and rarely getting into trouble themselves. They often managed to stay in the first notch of the reverser throughout, but with the main valve of the regulator in use whenever necessary. The more multiples of ten wagons on one train increased the risk of stalling because of the proportionally greater resistance especially on curves. There was also a dependence on up to three locomotives staying in shape for nearly half an hours hard steaming. The most ambitious train I saw in this respect, one cold November day in 1969, consisted of thirty-four loaded wagons.

Steve meticulously documented his journeys such as this one where loco *Stanley* is banked by loco *Repulse* on the journey from Astley Green Colliery to Walkden Yard:

Location	Time	Regulator	Cut-off	Boiler pressure (lbs/ sq in)	Average speed from last location
Astley Green	09.08	3/4	75%	160	15.5 mph
Booths Bank Crossing	09.13	2/3	75%	160	6 mph
East Lancs Rd Bridge	09.18	½	50%	150	8 mph
LNWR Bridge	09.21	½	25%	150	10 mph
L&YR Bridge	09.25	¼–2/3	25%	155	
App Walkden	09.26	½–¾	75%	150	
Walkden	09.29			140	9 1/2 mph from Astley

Steve recalled the memorable journey:

The trip on *Stanley* provided one of the most memorable highlights of my footplate experience. Words are inadequate to convey the sheer drama of the journey. From the first wagon, the internal, shovels full of coal were slamming into the cab backplate. The bunker was being filled on the move as the mechanical stoker banged and clonked away, emptying it at a similar rate. These sounds vied with the engine's roaring exhaust while noticeably screaming injectors played their part, along with a thunderstorm which raged around us. Inexplicable (but it is in my notes), large pieces of red hot coal were flying back into the cab via the firebox door, which if closed quickly became red hot itself. When we reached Walkden, *Stanley* took the train up to Ashton's Field in three single loads.

In mid-October Steve noted that locomotives *Francis* and *Bridgewater* had been partly cut up. In the workshops were *Respite* (for Astley Green) and *Warspite* (Hunslet 3776) from Harrington No. 10 Colliery, west of Lowca village, north of Whitehaven.

At times the workload was just too much, as Steve recalls after a visit to Walkden on Saturday 12 April 1969:

It was just as busy, however on 12 April, a really blustery day with the wind whipping up Hill Top Flash until it looked more like the sea! The first banker load (20 on) stalled on Walkden Bank after a 'flat out' grind at ¼ mph with *Warrior* leading and *Respite* at the back. During *Warrior*'s crew's dinner break, pies by the sand drier in Walkden Shed, Astley men worked another 18 wagons up using *Respite* banked by *Repulse*. The last train that day (total 86 wagons for the day) also conveyed 18, but double banked because one locomotive had to be left at Walkden for attention.

Steve occasionally recorded the sounds associated with the colliery locomotives (available today on CD) and recalls the following journey taken on Saturday 19 April 1969:

It rained heavily on the following Saturday, so I mostly rode on the footplate. It was an average day for traffic. *Warrior* slipped intermittently in the wet, but we got up satisfactorily on the first banked train. The second I recorded at Booths Bank Curve. *Repulse* really going at it at the back of 21 loaded plus an empty as the rain fell all around and plopped on my microphone. So the experiences went on, in all weathers, and always entertaining, but for how long?

On 18 April the *Bolton Evening News* had carried the shock headline 'SHUTDOWN BLOW FOR PIT HIT BY BAD LUCK'. Astley Green Colliery was to close on 27 June [1969], the tenth in the Leigh area to do so since 1955. The 1,427 workers were particularly worried about the announcement because of the threat to their livelihood. A series of accidents, mechanical breakdowns and a costly fire gave rise to the 'bad luck' tag. The NCB claimed that £600,000 had been lost in 1968, and area output targets for 1970 had been cut by about 1.5 million tons.

Steve carried on with his Saturday visits until hearing the news on 11 June that Astley Green Colliery had been given a reprieve until at least March 1970. Even three locomotives working together met their match in Steve's following account:

Things really ground to a halt on Saturday 29 November 1969. *Warrior* banked by *Respite* and *Harry* left Astley Green at 3.50 p.m. in a valiant attempt to move 34 wagons to Ashtons Field. Two thirds of them were filled with fine grade or slack, and so heavier than average (about 780

The woefully underpowered Yorkshire Engine Co. 0-6-0DE (2712/1958) at Ashton's Field coal blending plant, 8 April 1970. The strange structure behind was a road delivered coal tipping station. (Philip Hindley)

tons gross). Almost inevitably the train ground to a halt on Walkden Bank. With 600 tons pinned down on that precipitous slope and held by the two bankers, *Warrior* set off with the leading eight wagons and returned light soon after. The three locomotives were then able to restart the remaining twenty-six (still equal to about thirty) although not with ease and the day's total of ninety was safely deposited at Lynnyshaw Moss.

Steve recalls riding a Yorkshire Engine Co. 0-6-0 diesel-electric that had been brought into action to help clear the coal stocks on the ground at Astley Green Colliery after mining ceased on 3 April 1970. This locomotive had arrived in 1957 (works number 2660) to shunt in the Worsley area after locals made their feelings clear as regards the smoke! The Astley Green coal was destined for the refurbished coal blender at Ashton's Field. On 28 February 1970 Steve had a surprise, noting that the diesel required full power throughout the journey and took roughly the same time from Mosley Common to Ashton's Field although with 1½ wagons fewer. As Steve termed it, 'The diesel had a grindingly slow maximum speed' (of approximately 14 mph). The diesel's gearbox was to fail, taking it out of action by May 1970.

Steve ended his account as follows;

On Saturday 15 August [1970] all was dead and the steam locomotives were being washed out, but weekday evening activity through the middle of that month saw the re-boilered *Warrior* (not repainted) and *Stanley* sharing ordinary line traffic and three-wagon enthusiast specials, including the Branch Line Society, a true sign of the end. Sometimes both trains were in motion, with one a mere ¼ mile behind the other. That was the last rail traffic I saw, although movement continued officially until 2 October 1970.

Cliff Shepherd Witnesses the Final Days

Long-standing Industrial Railway Society member (in 2012 the publications editor) Cliff Shepherd visited the Central Railways system a number of times. He has kindly allowed me to reproduce three of his atmospheric photographs taken on 18 March 1970 from the Astley Green Colliery area, the pit ceasing mining coal shortly afterwards on 3 April 1970;

Looking towards Astley Green Colliery from the east on 18 March 1970, locomotive *Harry* is seen with fulls at the weighbridge. This is a superb record both of the end of a mining era and a mineral railway system. (Cliff Shepherd)

Long standing Astley Green Colliery locomotive *Stanley* leads loco *Respite* and thirty-three coal wagons to the north, 18 March 1970. On the distant horizon far left is Mosley Common Colliery winding house. The photo gives an idea of the incline to be tackled en route. (Cliff Shepherd)

Loco *Stanley* returns with empties from the exchange sidings at Astley Moss, south of the canal at Astley Green Colliery, 18 March 1970. A superbly evocative image by Cliff Shepherd.

Humphrey stood on death row at Walkden Yard for over seven years, and is seen here on 8 April 1970 minus coupling rod and smokebox door. Note the six secondary air inlets aside the smokebox door flange. Behind, an Austerity boiler stands on a flat wagon. (Philip Hindley)

A Century of Operation Closes: The Final Years

The old Bridgewater Collieries mineral railway network finally closed for coal transport by late 1970, a century and five months after the first locomotive was delivered to the Bridgewater Trustees. Walkden Yard staff no doubt were sad to hear the news but the yard still had a very important colliery engineering support role. The situation at the yard must have been relatively secure as it opened its doors to the public on Sunday 8 June 1975. In the 1970s Walkden even helped out with the restoration of locomotives on the Keighley & Worth

Astley Green Colliery had fully closed by August 1970. This photograph is annotated on the reverse in biro, 'Drivers and shunters. Astley Green. NCB at time of closure.' It probably has all the colliery loco staff in shot. From left to right: K. Redford, S. Dandy, R. Yates, K. Whittle, A. Flitcroft, T. Whittle, W. Hughes and R. Rylance. Behind the men is loco *Repulse*. (Jimmy Jones)

**NATIONAL COAL BOARD
CENTRAL WORKSHOPS – WALKDEN**

————

Open Day

SUNDAY, 8th JUNE, 1975

from

1.00 p.m. – 6.00 p.m.

————

**The Management of these Workshops
extend a welcome and invite you to see
behind the scenes of our industry.**

This brochure was printed when Walkden Yard opened its doors to the public in June 1975. A previous open day had been held in August 1966. By 1975 the colliery railway system had closed, Walkden concentrating on colliery engineering support work. The no expenses spared event featured a safety equipment exhibition, training facilities displays, recitals by Walkden Prize Band, a model colliery and equipment exhibition, a steam locomotive in operation, first aid demonstrations, historical photographs exhibition, trade displays of mining equipment, children's film shows, underground battery loco rides for children, a refreshments area, the ubiquitous tombola and a hobbies exhibition.

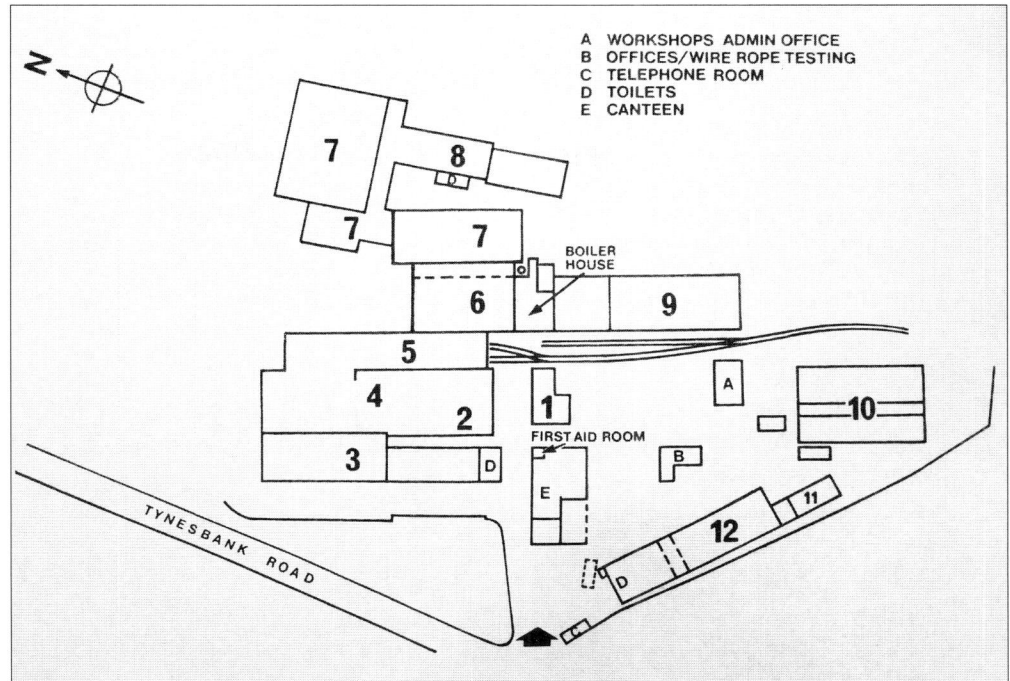

Open Day site map, June 1975. Compare this with the site plan of 1938 shown earlier. The key was as follows, giving us an idea of the work being carried out at that time.

1. **Hydraulic Pump Shop.** A variety of pumps are stripped in this shop and are rebuilt and tested before going back into colliery service to provide hydraulic power for coal face equipment.

2. **Blacksmith Shop.** Many different forged components are manufactured in this section for use at collieries and other departments in the workshops.

3. **Material Preparation Department.** Limited stocks of plate, bars and sections are stored here to meet the planned workshop needs; they are also cut, profile burned and pressed in this area to meet the requirements of other departments.

4. **Fabrication Shop.** This shop provides manufacture, repair and modification of components for powered supports work passing through workshops, together with the manufacture and repair of pit cages, skips [coal carrying cages], washery and screening plant and conveyor structure.

5. **Locomotive and Boiler Shop.** Complete overhaul of colliery surface and underground locos, i.e.; Battery, Diesel and Steam Locomotives. This section also carries out boiler retubing and liquid controller tube repairs along with repairs of sub-assemblies for colliery locomotives.

6. **Mechanical Department.** This is in three sections; machine section, fitting and turbine pump repairs. It handles a range of work through the many machine tools installed, plus repairs to air compressors and about forty various makes and sizes of pumps.

7. **Powered Support Departments.** This is in three sections; namely stripping, cleaning and assembly shops for all components in the repair, overhaul and test of both Gullick, Dobson and Dowty type self advancing roof supports for mechanised coal faces.

8. **Cage Suspension Shop.** The cage suspension department carry out routine examination of suspension gear, rope, capels, detaching hooks and man-riding gear [transport systems below ground].

9. **Cable Shop.** Damaged cables are received direct from collieries and the work of the department is to locate faults, repair, test and return to the pit. New cables are also made up.

10. **Valve and Leg Repair Shops.** Valves and legs [from underground coal face hydraulic roof supports] are stripped, cleaned and inspected prior to repair and tested before being returned to the powered support assembly section or colliery service exchange.

11. **Tank Shop.** A variety of tanks are stripped, repaired and rebuilt in this section before being returned to colliery service for use with underground hydraulic equipment.

12. **Joiners Shop.** The main services include manufacture of cable drums, air doors [ventilation doors below ground] and frames and Walkden-type flushing shields for colliery requirements and powered support installations [used on coal face equipment to stop roof-fall debris reaching men on the coal face working area] from scrap PVC belting supplied from collieries, together with any colliery joinery requirements. Also included in this section are painters, bricklayers and a plumber.

Valley Railway, Steamport Southport and the former Dinting Railway Centre near Glossop (closed 1991). Between October 1973 and January 1974, oil prices quadrupled – coal suddenly seemed worth investing in once more and yet another version of the long line of 'Plan for Coal' strategies was rolled out, this being a ten-year plan. The plan required an investment of £1,400 million to increase the industry's capacity by sinking new collieries, extending existing collieries and increasing opencast mining production. The industry had renewed confidence for the future, the wider public needed to be informed of the plans and a PR exercise unfolded; visits to collieries were organised. At Walkden Yard an open day was organised in June 1975 (see illustrations).

Walkden Yard in the Late 1970s and 1980s, the Range of Work

Walkden carried on repairing locomotives (as well as maintaining all manner of mining equipment, which it always had) for the NCB until 1980. The last loco to be repaired at Walkden was a Rolls-Royce-engined 650-hp Sentinel diesel, returning back to Sutton Manor Colliery, St Helens, in October 1983 shortly before the start of the 1984–85 miners' strike. The colliery closed in 1991.

The miners' strike was problematic for Walkden Yard staff who were members of various mining-related unions but not particularly fervent supporters of the action. The announcement of the closure of the Kirkless Workshops, Wigan, by area management in Stoke-on-Trent was made in mid-June 1985, 300 jobs being lost by October. Some men took redundancy, others transferred to Trentham Workshops in Stoke, and others headed to Walkden Yard. The desperate state of affairs post-strike can be seen by the fact that area NUM officials at Bolton demanded that only NUM-affiliated men should be allowed to transfer to Walkden, concerned that the arrival of a small number of breakaway union-organising members from Kirkless might lead to a foothold at Walkden. Sadly the power of the unions had been crushed by the strike and area management could do as they pleased with colliery workshops, which were being closed down nationally with the forced run down of the coal industry

The official closure announcement (rather than a proposal open for negotiation, as might have been the case pre-strike) for Walkden Yard came from the recently-created British Coal on 13 August 1986 along with Ansley Workshops near Nuneaton. 250 men lost their jobs at Walkden, 160 at Ansley. Work on-site at Walkden had ceased totally by November 1986.

As curator of the Lancashire Mining Museum at Buile Hill Park, Salford, I paid many visits to Walkden Yard to photograph the site, to collect any historical documentation liable to be lost and importantly was able to persuade the manager (after a guided tour of Astley Green Colliery museum!) to allow members of the Astley group to collect large amounts of tools and general equipment for their workshops. I remember in the castings pattern store seeing wooden formers of loco parts, a certain number of these heading to Astley Green. All this thankfully took place before the usual Area Raiding Party from Chatterley Whitfield Mining Museum, Stoke, arrived. Chatterley Whitfield Museum closed in August 1991; sadly many objects controversially taken to the museum from the closing Lancashire collieries were auctioned off and headed off in all directions.

Diverting slightly, the Worsley underground canal, which had been accessible at Ellesmere Colliery adjacent to Walkden Yard, ceased coal transport after 1887. It then served as a gigantic mining sough, draining water from old workings of the Bridgewater complex as well as the strata that the system passed through. The last mine water actively pumped into it was in 1968, the year of the closure of the last former Bridgewater pit, Mosley Common. The last NCB inspection of the main canal level from Ellesmere Pit, Walkden, (adjacent to Walkden Yard) to Worsley Delph took place on 28 September 1968.

It is reassuring that a few years after Walkden Yard was demolished a substantial history of the Lancashire colliery railways was gradually produced by four fine researchers and rail historians: C. H. A. Townley, C. A. Appleton, F. D. Smith and J. Peden in *The Industrial Railways of Bolton Bury and the Manchester Coalfield*. Sadly none of the authors is now with us, but their work is rightly regarded as the definitive study on the subject so far.

Houses now stand on the site of Walkden Yard, which was for centuries farmland and later a hive of mining activity, accommodating Ellesmere Colliery then Walkden Yard itself. Bridgewater Estates itself survived until its purchase by the giant land and property development concern Peel Holdings in 1984, retaining its name until 1987.

Farewell to Walkden Yard

Walkden Yard after closure, 1987. From the left: Cable Shop; Boiler House chimney; Mechanical Dept; Powered Supports Dept; and the old loco shed, now the Transformer Shop. (Alan Davies)

Walkden Yard after closure, 1987. From the left: Rope Testing; Offices; Loco Shop; Mechanical Dept; Boiler House chimney; Cable Shop. Staffa type hydraulic mining motors are stacked on the racking. (Alan Davies)

Walkden Yard after closure, 1987. This photo was taken from close to the main entrance, with the former Blacksmith Shop in the middle distance, along with the Boiler House chimney and the Fire Station. The building to the far right was (I am told) a canteen in the 1960s. (Alan Davies)

Walkden yard after closure, 1987. From the left: Fabrication and Pump Testing; Rope Testing; Loco Shop. The buildings on the right were (I am told) fitting shops in the 1960s. (Alan Davies)

Walkden Yard after closure, 1987, viewed from the northern Manchester Road end of the site. Buildings in use at the time were, from left: Powered Supports; Mechanical Dept; Loco and Boiler Shop; and the Fabrication Shop. In the foreground are mining roof support canopies, on the right Gullick hydraulic pumps on skids. (Alan Davies)

Appendices

Appendix 1
Extract from an Agreement Between Manchester Collieries Ltd and the Central Wagon Co. Ltd. for the Sale of Railway Wagons, Dated 3 April 1945. (Courtesy of and transcribed by Philip Hindley)

Whereas Manchester Collieries Ltd. own 8398 wagons and have hired 1150 wagons under hire purchase agreements and have agreed to sell the wagons and their rights under the hire purchase agreements to Central Wagon Co. Ltd for the sum of £585,110 of which £526,387 shall be payable in cash and the balance of £58,723 shall represent the outstanding hire purchase liabilities to be taken over by Central Wagon Co. Ltd."

Manchester Collieries Ltd. to pay compensation if:
(1) The wagons or any of them are compulsorily acquired within ten years of 31/3/1945 under any Act of Parliament or statutory rule or order in council.
(2) Any wagons are condemned or excluded from main line traffic during a period of ten years from 31/3/1945
Manchester Collieries Ltd. to hire the wagons from Central Wagon Co. Ltd. from 1/4/1945 and will (among other things) keep wagons in good repair in running order and do all necessary greasing and oiling to axles and axle trees.

As soon as practicable after the wagons cease to be requisitioned Manchester Collieries Ltd. shall cease to repair main line wagons, thereafter all the wagons to be repaired by Central Wagon Co. Ltd. at Manchester Collieries Ltd. expense.

First Schedule – Hire Purchase Agreements (all 10 years)

Wagon Finance Corporation Ltd.	200	31/5/1936 (date commenced)
Wagon Finance Corporation Ltd.	50	4/6/1935
Wagon Finance Corporation Ltd.	50	1/12/1936
Wagon Finance Corporation Ltd.	175	1/1/1937
Wagon Finance Corporation Ltd.	150	1/9/1939
Lancashire & Yorkshire Wagon Co. Ltd.	75	12/11/1936
Lancashire & Yorkshire Wagon Co. Ltd.	150	15/9/1939
Chas. Roberts & Co. Ltd.	100	1/12/1936
Chas. Roberts & Co. Ltd.	200	1/10/1939

2nd Schedule – Valuations

Capacity	Year Built	Cost Price	Central Wagon Valuation
8T	All	£30	£20
10T	All	£40	£35
12T	Before 1920	£50	£50
12T	1923/4	£100	£70
12T	1925–7	£110	£75
12T	1928	£110	£80
12T	1934–7	£140	£100
12T	1939	£150	£130

3rd Schedule – Rate of hire per week

8T	All	1s 6d
10T	All	2s 0d
12T	Before 1920	3s 0d
12T	1920–30	3s 6d
12T	After 1930	4s 3d

Agreement Between Manchester Collieries Ltd and the Central Wagon Co. Ltd and the Central Wagon Hiring Co. Ltd Dated 1 April 1946

Included:
(1) Number of new wagons subject to 3/4/1945 Agreement is 9530
(2) 9530 wagons sold by Central Wagon Co. Ltd. to Central Wagon Hiring Co. Ltd.

(3) All obligations of Manchester Collieries Ltd. under the earlier agreement "should be and become and should endure" (*quote)* for the benefit of the Central Wagon Hiring Co. Ltd and its assigns in lieu of the Central Wagon Co. Ltd.

Agreement Between Manchester Collieries Ltd and the Central Wagon Hiring Co. Ltd Dated 24 June 1946

Covered the hire of 9515 wagons for 9 years from 1/4/1946 and included the same figures for valuations and rates of hire as the agreement of 3/4/1945

Appendix 2

A selection of siding plans (overleaf) of the early 1960s (possibly 1961) produced by the NCB to show the position of weigh machines. The originals have alterations scribbled on them, the ever changing tipping lines at Cutacre for example, but they may be of use to modellers.

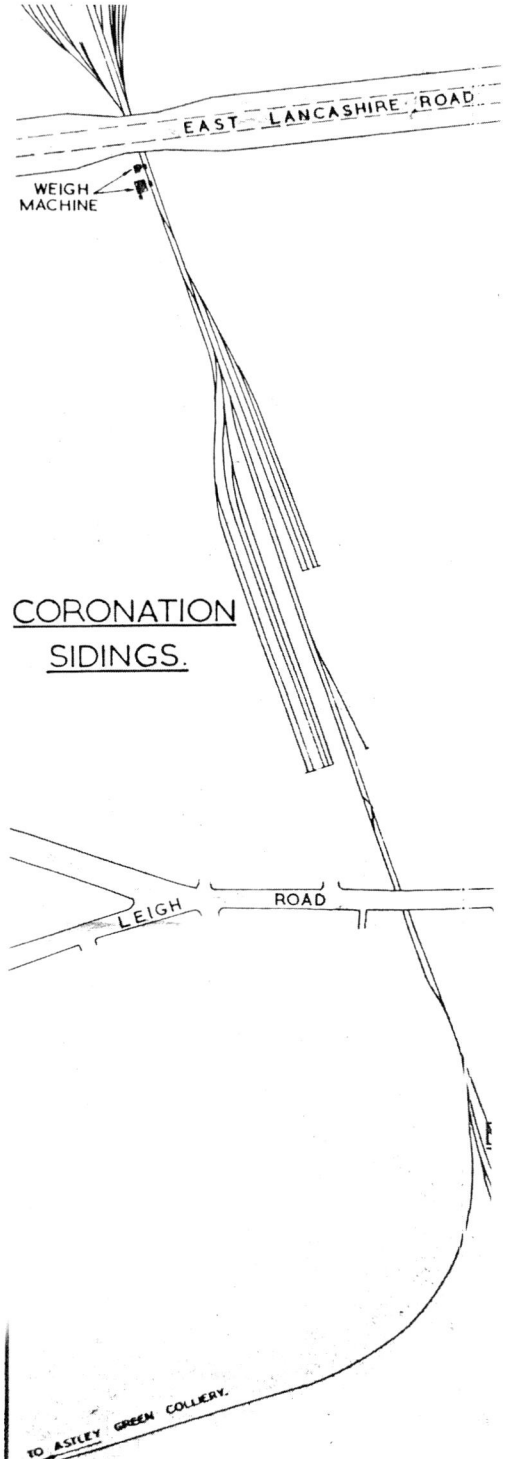

EAST LANCASHIRE ROAD

WEIGH MACHINE

CORONATION SIDINGS.

LEIGH   ROAD

TO ASTLEY GREEN COLLIERY.

Coronation sidings, south of the East Lancs Road, named so after 2 June 1953.

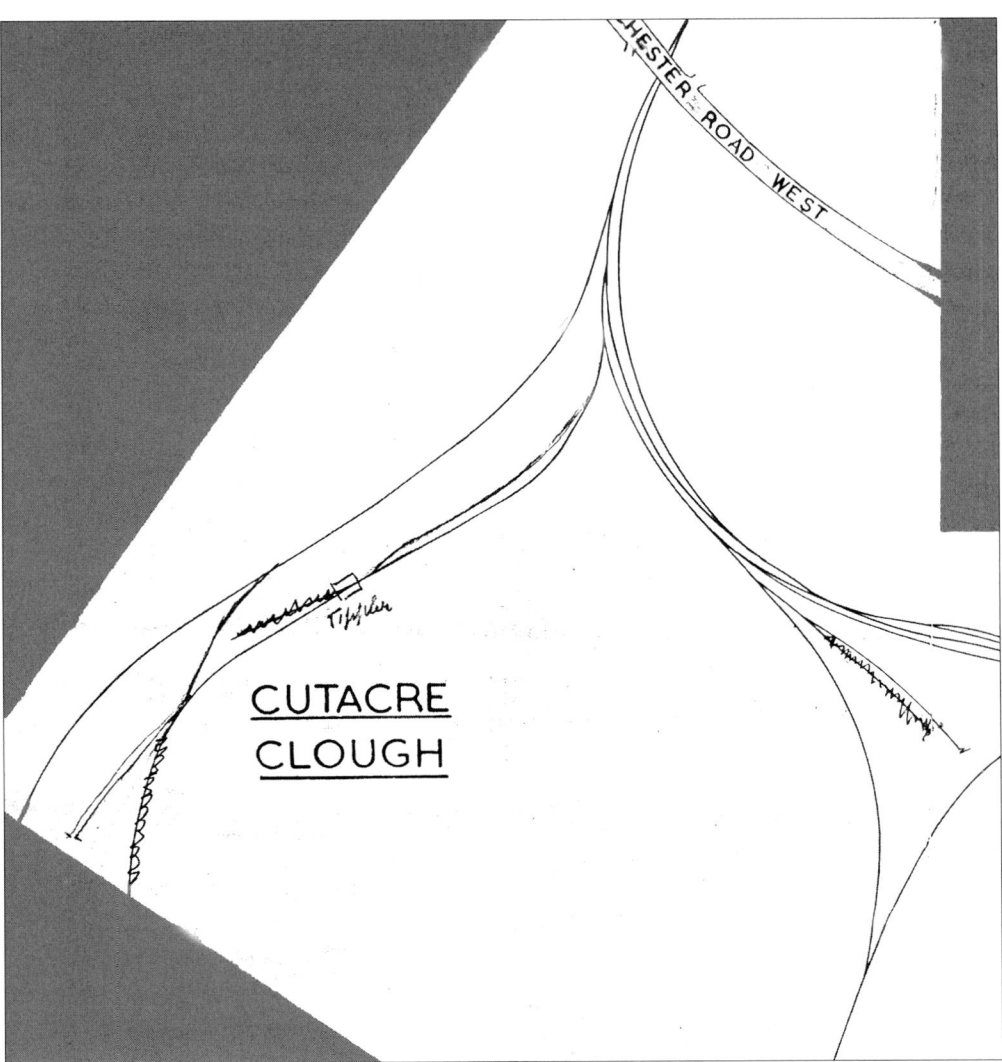

Lines to the massive Cutacre waste tip, south of Brackley Colliery, east of Little Hulton. The tip eventually made its way into the *Guinness Book of Records* as the largest in Britain.

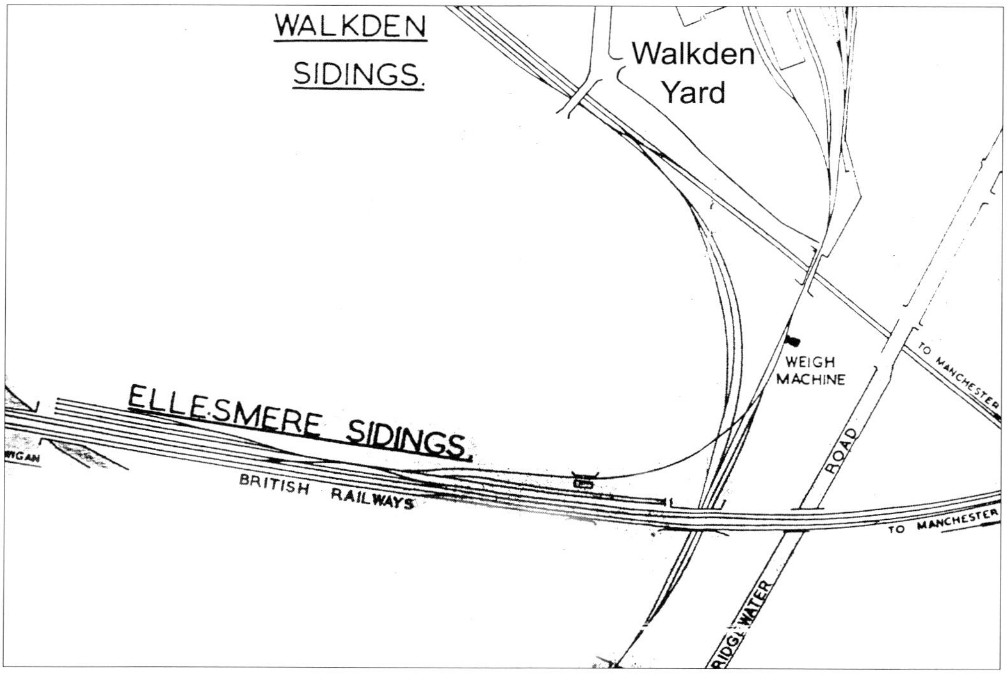

WALKDEN
SIDINGS.

Walkden
Yard

WEIGH
MACHINE

TO MANCHESTER

ELLESMERE SIDINGS.

WIGAN

BRITISH RAILWAYS

ROAD

TO MANCHESTER

RIDGE WATER

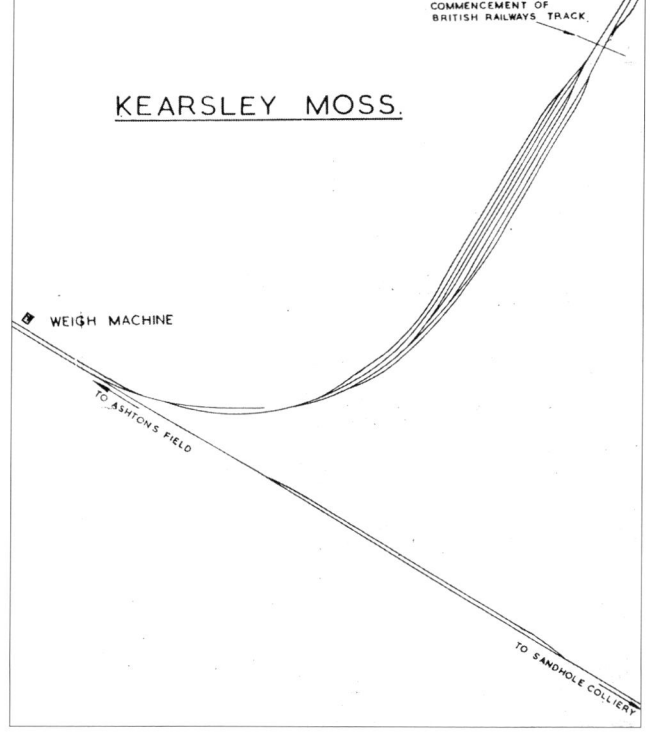

COMMENCEMENT OF
BRITISH RAILWAYS TRACK.

KEARSLEY MOSS.

WEIGH MACHINE

TO ASHTONS FIELD

TO SANDHOLE COLLIERY

Above: Lines south of
Walkden Yard showing the
location of both Ellesmere
and Walkden sidings.

Right: The BR exchange
sidings on Kearsley
Moss, the Ashton's Field–
Sandhole line below.

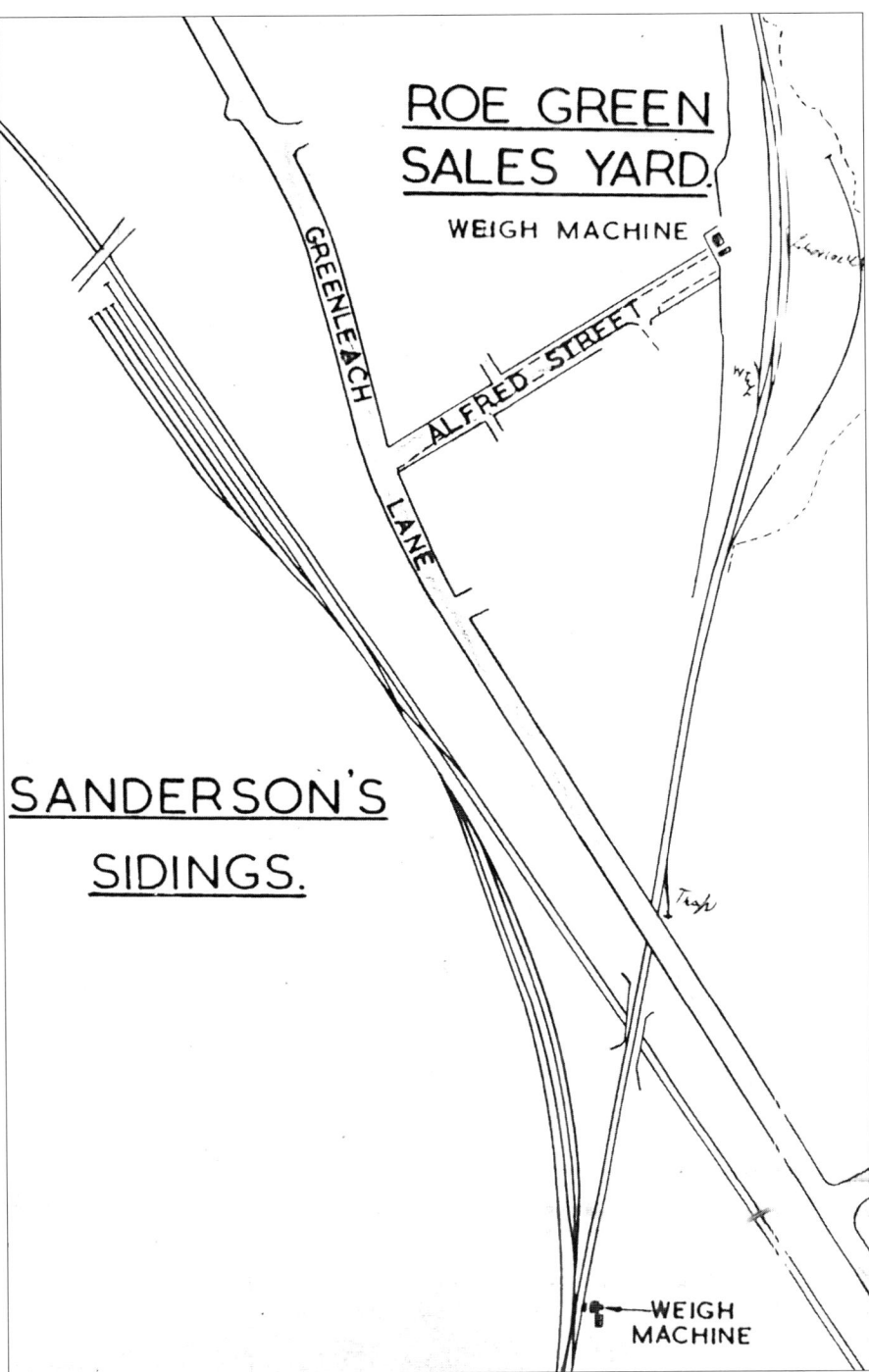

Sandersons sidings, Worsley, a popular vantage point for rail enthusiasts and photographers over the years.

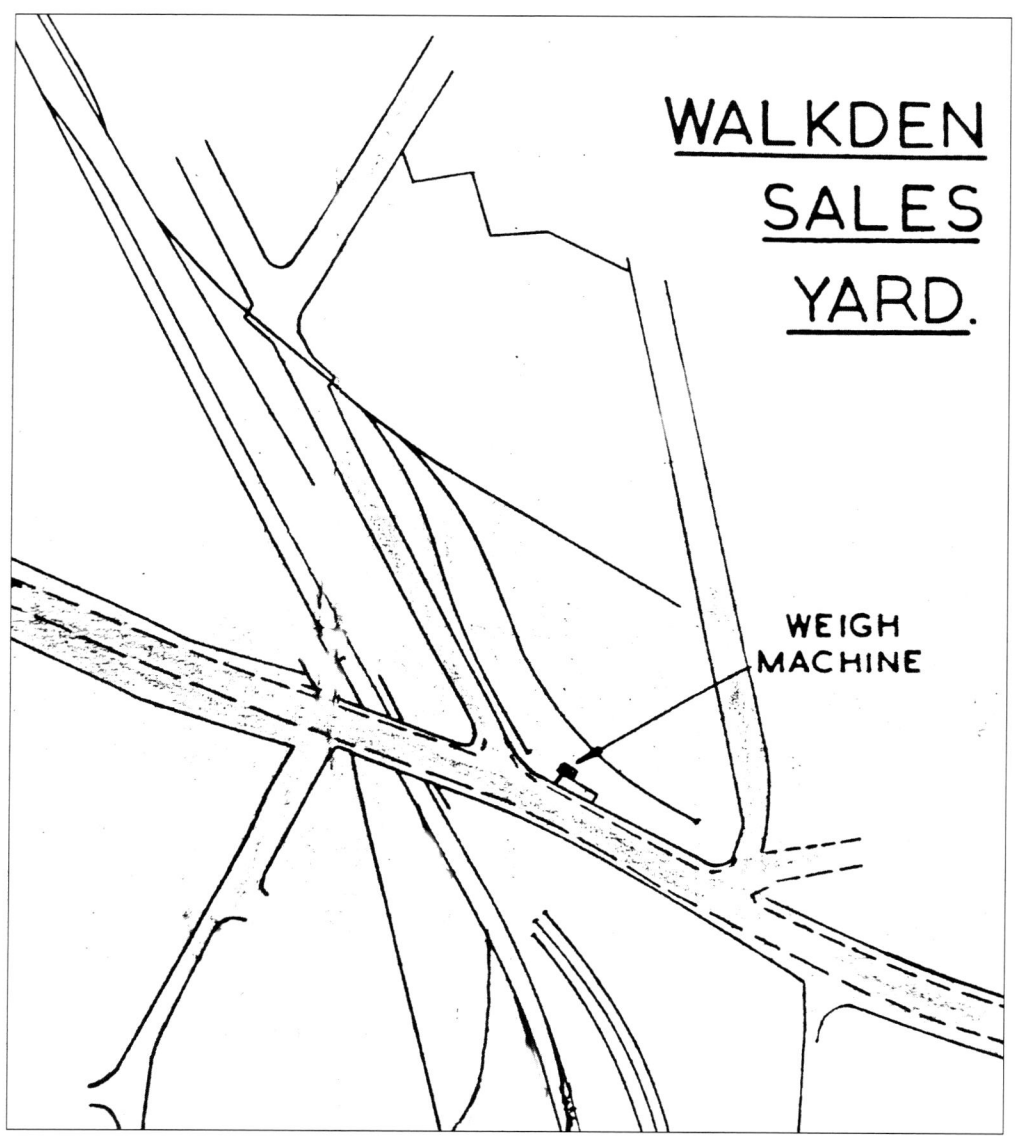

Walkden coal landsale yard operated where the first Walkden Yard had been situated and survived as such until the 1970s.

Bibliography and Sources

Anderson, Donald, *Coal – A Pictorial History of the British Coal Industry* (Newton Abbot: David & Charles, 1982).

Anderson, Donald and J. Lane, *Mines and Miners of South Lancashire 1870–1950* (Donald Anderson, 1980).

Arnot, R., *The Miners, In Crisis and War from 1930 Onwards* (Allen & Unwin Ltd).

Ashworth, W., *The History of the British Coal Mining Industry. Vol. 5 1946–1982* (Oxford University Press, 1986).

Booth, G. L., *The Trustees Railways and After* (1994).

Booth, George L., *Walkden Yard 1932–1939* (Walkden Library Local History Group).

Booth, George L., *Walkden Yard Yarns No. 1*.

Booth, George L., *Walkden Yard Yarns No. 2*.

Bradley, V. J. and P. G. Hindley, *Industrial Locomotives of Lancashire: Part A, The National Coal Board Including Opencast Disposal Points and British Coal* (Industrial Railway Society, 2000).

Brooks, E. A., *Working On't Railroad 1939–1950* (1991).

Chapleon, A., *La Locomotive à Vapeur* (Translated by G. Carpenter, 2000).

Church, R., *The History of the British Coal Mining Industry. Vol. 3 1830–1913* (Oxford University Press, 1986).

Colliery Guardian.

Colliery Year Book & Coal Trades Directories (1943–1963).

Davies, Alan, *The Wigan Coalfield* (Tempus, 1999–2009).

Davies, Alan, *The Pit Brow Women of the Wigan Coalfield* (Tempus, 2006).

Davies, Alan, *The Atherton Collieries* (Amberley, 2009).

Davies, Alan, *The Pretoria Pit Disaster, A Centenary Account* (Amberley, 2010).

Davies, Alan, *Coalmining in Lancashire & Cheshire* (Amberley, 2010).

Ezra, Sir Derek and Others, *Coal – Technology for Britain's Future* (London:

Macmillan, 1976). Flinn, M. W., *The History of the British Coal Mining Industry. Vol. 2 1700–1830* (Oxford University Press, 1984).

Frost, Roy, *Electricity in Manchester* (Neil Richardson, 1993).

'Guide to the Coalfields' in *Colliery Guardian* (1948–1986).

Griffin, Dr A. R., *Coal Mining* (Longman, 1971).

Hayes, Geoffrey, *Collieries & Their Railways in the Manchester Coalfields – 2nd Edition* (Landmark, 2004).

Industrial Railway Record

Lewis, B., *Coal mining in the Eighteenth and Nineteenth Centuries* (London: Longman, 1971).

Manchester Collieries Ltd Newsletter (1944, 1945, 1946).

NCB/Bridgewater Collieries records (Lancashire Record Office, Preston):

NCB Bw 11/1–2, *Valuation of Bridgewater Collieries, etc., for estate duty on the Third Earl of Ellesmere (deceased) as on 13th July 1914.*

NCB Bw 11/5, *Bridgewater Collieries in Reconstruction as at 31st December 1920 (valuation done as additions or deductions to valuation of 13/7/1914).*

NCB Bw 19/1, *General correspondence, papers and plans of John and James Ridyard, colliery agents, Walkden Moor 1807–1867.*

NCB Bw 20/15, *Bridgewater Collieries Ltd. Heavy Expenditure Book 1921–28.*

NCB Bw 20/16, *Property ledger. Valuations of structures, plant, etc. 1921–1930.*

NCB Coal Magazine (1947–1960)

Nef, J. U., *The Rise of the British Coal Industry* (Routledge, 1952).

North Western Coalfields Regional Survey Report (Ministry of Fuel & Power, HMSO, 1945).

Pamely, C, *The Colliery Manager's Handbook* (Crosby Lockwood, c. 1904).

Porta, L. D., *Steam Locomotive Power: Advances Made During the Last Thirty Years. The Future.* (1990 and 1991).

Preece, G., *Coalmining* (City of Salford Cultural Services Dept, 1981).

Railways Illustrated. No. 61.

Railway Magazine (August 1963).

Railway Magazine (January 1990).

Railway Magazine (February 1992).

Record (Industrial Railway Society):

Hayes, G., No. 44 *Joseph* and *Bridgewater* article with photographs.

Golding, Cyril, No. 172 *Destination Walkden Yard.*

Hindley, P. G., D. Holroyde and S. Oakden, No. 196 *Modifications to NCB Steam Locomotives* (includes details of modifications carried out at Walkden to reduce smoke)

Leyland, Steve, No. 172 *Walkden: The Last Two Years.*

Redmayne, R. A. S., *The Problem of the Coal Mines* (Taylor & Francis, 1945).

Supple, Prof. B., *The History of the British Coal Mining Industry. Vol. 4 1913–1946* (Oxford University Press, 1987).

The Industrial Locomotive (Industrial Locomotive Society):

Appleton, C. A., No. 13 *Recollection of the Bridgewater Section of Manchester Collieries.*

Appleton, C. A., No. 41 *Austerity Tanks at Collieries in Lancashire.*

Appleton, C. A., No. 42 *Austerity Tanks at Collieries in Lancashire – Part 2.*

Appleton, C. A., No. 43 *Austerity Tanks at Collieries in Lancashire – Part 3.*

Appleton, C. A., No. 45 *Austerity Tanks at Collieries in Lancashire.*

Appleton, C. A., No. 47 *Austerity Tanks at Collieries in Lancashire.*

Appleton, C. A., No. 60 Katharine *article with photo.*

Appleton, C. A., No. 62 *From Walkden Yard Records Part One.*

Appleton, C. A., No. 63 *From Walkden Yard Records Part Two.*

Appleton, C. A., No. 66 *Accident at Wharton Hall.*

Appleton, C. A., No. 67 William *of Walkden.*

Appleton, C. A., No. 72 *Another Adventure at Wharton Hall.*

Appleton, C. A., No. 77 *The Wanderings of* Robin Hood.

Appleton, C. A., No. 79 *The Wanderings of* Carbon.

No. 67 Katharine *works photo.*

The Locomotive Magazine (15 October 1924).

The Railway Magazine (March 1963).

Townley, Appleton, Peden, Smith, *The Industrial Railways of the Wigan Coalfield Series, Parts 1 and 2: The Manchester Coalfield, Bolton & Bury* (Runpast, 1994 and 1995).

Transactions of the Institute of Mining Engineers.

Townsley, D. H., *The Hunslet Engine Works* (1998).